The SCIENCE *of* LIFE

ALSO BY S.A. BARNETT

The Human Species
'Instinct' and 'Intelligence'
The Rat: a Study in Behavior
Modern Ethology: the Science of Animal Behavior
Biology and Freedom

The SCIENCE of LIFE

FROM CELLS TO SURVIVAL

S. Anthony Barnett

ALLEN & UNWIN

Copyright © S. Anthony Barnett 1998

All rights reserved. No part of this book may be reproduced or transmitted in any form or by any means, electronic or mechanical, including photocopying, recording or by any information storage and retrieval system, without prior permission in writing from the publisher.

First published in 1998
Allen & Unwin
9 Atchison Street, St Leonards 2065 Australia
Phone: (61 2) 9901 4408
Fax: (61 2) 9906 2218
E-mail: frontdesk@allen-unwin.com.au
Web: http://www.allen-unwin.com.au

National Library of Australia
Cataloguing-in-Publication entry:

Barnett, S. A. (Samuel Anthony), 1915– .
    The science of life.

    Includes index.
    ISBN 1 86448 610 4.

    1. Biology—Popular works. 2. Human biology—Popular works.
    3. Evolution (Biology)—Popular works. 4. Human ecology—Popular
    works. I. Title.

599.9

Set in 11/13.5 pt Janson Text by DOCUPRO, Sydney
Printed and bound by South Wind Production (S) Pte Ltd, Singapore

10 9 8 7 6 5 4 3 2 1

Biologists work very close to the frontier between bewilderment and understanding. Biology is complex, messy and richly various, like real life.

    P.B. MEDAWAR

CONTENTS

Preface — ix
Acknowledgement — xi

IN THE BEGINNING

1 The Earthbound Phenomenon — 3
2 Evolution: History or Blasphemy? — 12

THE UNSEEN WORLD

3 Cells, Sex and DNA — 31
4 The Authentic Gene — 58
5 The Smallest Organisms: Matters of Life and Death — 87

THE IMPORTANCE OF BEING CONSTANT: BODIES AND MINDS

6 The Stability of the Body — 99
7 Brains and Mind — 124

ENVIRONMENTS

8 Ecology: Species Living Together — 147
9 Social Lives — 174

A STRUGGLE FOR EXISTENCE?

10 The New Darwinism   195

THE HUMAN SPECIES

11 Human Nature   219
12 Humanity in Nature   237

Appendix I   A Classification of Organisms With
              Miscellaneous Annotations   253
Appendix II   A Stratification of Organisms: The Record
              of the Rocks   265
Appendix III Glossary   268
Further Reading   283
Index   287

PREFACE

A FAMOUS PRODUCER OF MOVIES, Samuel Goldwyn, once demanded a story that begins with an earthquake and works its way up to a climax. Our story opens with the origin of life, in conditions much more violent than those of a mere earthquake. The climax is the appearance of humanity. The human species has, by itself, already drastically altered the face of nature. We can be confident, very soon, of still more extensive changes, some welcome, some frightening.

This book is about the nature of life and our place in the biosphere. Many of the most exciting recent findings in biology have been in genetics, cell physiology and biochemistry. 'Engineering' the genes and immunology are beginning to have sensational applications in economic biology and medicine. But much of the book is about whole organisms, especially what they do to us and what we do to them.

Biological science saves lives, improves health, grows more food, preserves nature and provides pleasure. It also presents problems. Does an emphasis on ecology make the author an ecofreak—or even an ecoterrorist? Should physicians treat the body as a machine and teachers regard their pupils' brains as computers? Are people only apes? Are apes (almost) human?

Are women fit by their biology only for the kitchen and the nursery? Is a gene in a test tube the equivalent of a genie in a bottle? What should we make of the fashionable misanthropy which presents human beings as incurably nasty because of their evolution?

In addition, every householder is daily faced with biological questions. The statements in advertisements on behalf of foods, drinks and other goods are not reliable guides to daily living: knowledge about how our bodies work is needed if they are to be interpreted safely; without it, words such as vitamins and even energy become no more than a magician's abracadabra.

To cope with everyday life, to enjoy our natural surroundings and to understand ourselves, we need to know something of the living world and of modern biology.

ACKNOWLEDGEMENT

MANY KIND PEOPLE HAVE helped me in conversation, in correspondence or by commenting on fragments in draft form. I am grateful to all of them. But my greatest debt is to the uncounted scientific workers whose labours have created the modern science of life. Very few are well known. In a famous lecture, given in 1923, J.B.S. Haldane described biologists as at first sight only poor, scrubby, underpaid persons, for whom the central problem of life 'may be the relationship between the echinoderms and brachiopods and the attempt to live on their salaries'. Nonetheless, he says, 'the biologist is the most romantic figure on earth at the present day'. Haldane's statement, which perhaps startled his audience, was in advance of his time. But now biology and biologists are in the centre of the global stage.

S.A.B.
June 1997

The biosphere. A view of an African tropical forest. Thousands of interacting species are present in this small area. The magnificent tree in the foreground, with buttress roots, is *Musanga cecropioides*. At night, in photographs taken from satellites, the burning of such forests sometimes shows up more brightly than the lights of cities.

IN THE BEGINNING

What is life? How did it arise? The fundamental idea of organic evolution still causes argument.

Chapter 1

The Earthbound Phenomenon

> The Greeks said that to marvel is the beginning of knowledge and that where we cease to marvel we may be in danger of ceasing to know.
>
> E.H. Gombrich

By 1989, the first grand tour of the solar system had been completed. The spacecraft, *Voyager*, had visited even Neptune, the most remote of the four great planets. As a result, we realise, more clearly than before, that our own planet is—with minor possible exceptions—the only known celestial body capable of giving a home to living things. If other, far distant solar systems allow life, we cannot detect it. On this, writers of science fiction hold the field.

The Biosphere

In a famous example of such fiction, *The Black Cloud*, by a distinguished English astronomer, Fred Hoyle, Earth is faced with disaster by the approach of an enormous mass of gas. The Cloud, however, proves to be intelligent. The hero of the story (a distinguished English astronomer) manages to communicate with it; and the Cloud sheers off.

The living things on earth, encountered by the Cloud, make an extremely thin, ragged film, the biosphere, spread on a small, cool fragment of matter, our planet. The unusual temperature of the earth's surface allows the presence of large amounts of water in the liquid state. All active organisms consist largely of water. The human body, with 60 or more per cent, is quite typical.

Throughout the biosphere, the substances characteristic of living things, or their products, are exchanged among organisms: animals such as sheep live on the carbohydrates, proteins and other substances in plants; others, such as tigers, live on sheep. Plants live partly on the products of decayed animals (including tigers and sheep) and of other plants. From this point of view, the biosphere is a unit. Nearly everything in it is recycled.

Each type of being, however, imposes its own kind of organisation on the material it takes in. When an animal has eaten, it digests large molecules to smaller ones; then the smaller units are reassembled in different larger ones. A plant takes in simple substances (especially carbon dioxide, nitrates and water) and synthesises complex ones. All organisms construct proteins typical of their own species or variety. An organism is therefore a special kind of system—a dissipative structure. It may exist for a long time as recognisably the same object, yet many of the atoms of which it is composed are constantly changing. A flame too, such as that made by lighting a steady gas jet, is a dissipative structure but an organism differs from a flame in its immense complexity: in particular, it regulates its own changes of energy; an example of such regulation is eating when one is hungry.

The chemical changes, or metabolism, which go on while an organism transforms and makes use of its nutrients, require special conditions. Most living things are active only in a narrow range of temperatures. Plants grow between the freezing point of water and about 50 degrees Celsius. Below zero, if they survive, they become inert; at higher temperatures proteins are destroyed. Most animals function only within a similar range. The obvious exceptions are well insulated birds and mammals in environments below zero; these produce heat internally and so keep their tissues at or near 40°C— another example of self-regulation. Exceptions are also found among bacteria which survive astonishing extremes of heat and cold.

Other restrictions are chemical. Organisms usually live either in or near water. The primary home of life is the sea, which contains about 3.5 per cent common salt and many other substances necessary for life. The tissues of land animals are permanently bathed in a salty solution, of almost constant composition, rather like that of the sea water in which their ancestors lived. Is this a chemical nostalgia for the remote past? Doubtful: the similarity is not very close. Salty body fluids do, however, provide a kind of marine hospitality for diverse modern organisms, from bacteria to worms.

Living things also need protection from electromagnetic radiation. Two kinds are especially important. Ultraviolet (UV) radiation is a source of sunburn and, after long exposure, of cancer. It is present wherever the sun shines; but much UV is absorbed, before it reaches the earth's surface, by a high concentration of ozone in a layer around 20 kilometres above sea level. Today this fragile shield is being eroded by fluorocarbons and other substances produced by human action. One likely outcome is an increase in a form of skin cancer (malignant melanoma). Large scale consequences, for instance ill effects of extra UV radiation on plant growth, also now seem possible.

The second kind, ionising radiation, includes the X-rays from machines and the gamma rays from nuclear fission. The destructive effects of X-rays on human tissues are rigorously guarded against when they are used medically. In the biosphere, most gamma rays come from naturally occurring radioactive elements and are present only at low intensity. High intensities arise, again, from human action. The injuries they can cause among human beings have been hideously demonstrated by the consequences of the atomic bombs dropped, in 1945, on two Japanese cities, Hiroshima and Nagasaki. The immediate effects of high doses were radiation sickness and death. Lower doses later resulted in leukemia and many other cancers. Probably, uncounted early abortions occurred. Other effects on unborn children led to severe mental retardation. Radiation also damages the genetical material (chromosomes and genes) of all organisms. As a result, late in the twentieth century, gross abnormalities are still appearing among the descendants of victims who survived the bombing.

OMNE VIVUM EX VIVO: ALL LIFE FROM OTHER LIFE

The chemical composition of living things, including ourselves, is known with great accuracy but nobody has yet created new life by assembling the necessary elements in a test tube. The rule is: organisms arise only from other organisms, each kind 'after its own kind'. This is new knowledge. Until the eighteenth century, spontaneous generation of life from not life, or abiogenesis, was usually taken for granted even by the learned. Maggots, it was supposed, grew out of dead meat; weevils obviously, it seemed, were generated by stored grains. The great chemist and Capuchin friar, J.B. van Helmont (1577–1644), who invented the word gas, was convinced that he had observed the generation of rats from bran and old rags. Such presumptions are still held by some people, especially warehousekeepers whose defective hygiene has allowed the breeding of pests in stored foods.

The change in outlook went with the rise of experimental methods. A Florentine physician and poet, Francesco Redi (1628–1698), who was one of the founders of modern parasitology, allowed meat to rot in vessels covered with thin cloth which admitted air but no flies. The 'control experiment' was to expose similar portions of meat, in similar containers, without the cloth. The covered meat putrefied but generated no insect larvae. Flies visited the unprotected meat and maggots duly developed from eggs laid by the flies. The reader could, at risk of some unpopularity, easily repeat this experiment.

Redi did not, however, settle the question of abiogenesis. What of the microscopic organisms which appear in all decaying matter? The French genius, Louis Pasteur (1822–1895), performed the decisive experiments in 1860. As a person, Pasteur sounds like a fictional scientist—dedicated, single-minded, autocratic and humourless. On a wedding anniversary, his wife wrote that he was busy as always: he 'says little to me, sleeps little, rises at dawn—in a word, continues the life that I began with him 35 years ago today'.

As a young man, Pasteur was a school teacher in Strasbourg, where he made a fundamental contribution to the study of crystals and of molecular structure. When, in 1855, he became a professor of chemistry, he turned to the study of fermentation—the process which produces alcohol in wine, beer and other beverages. This, he said, was due to microorganisms which had infected the liquid from outside. Similarly, putrefaction was not 'spontaneous' but was due to infection with microbes of decay. Opponents, however, were convinced that the organisms in wine and beer arose as a *result* of fermenting and they did ingenious experiments to prove it.

Pasteur refuted spontaneous generation by elaborating on and improving the methods of his opponents. He boiled and cooled solutions of sugar in glass vessels. To some he added dust particles, acquired by filtering air through cotton wool. He then sealed the flasks. The solutions grew the expected microorganisms. Next he boiled sugar solution and sealed the necks of the flasks while it was still boiling. In these uncontaminated solutions no growth followed. Pasteur also admitted air to flasks at high altitude among the glaciers of the Swiss mountains. These too grew nothing. He concluded (what we now accept) that the microscopic bacteria, yeasts and other organisms, which grow in such liquids, are descended from airborne organisms of the same kind: yeasts arise from other yeasts and so on.

Pasteur's conclusions had major commercial applications. Even in his own time they helped the producers or vendors of wine, beer and cheese: all require the presence of particular organisms during production; other organisms must be kept out.

Still more important were his contributions to medicine. Many illnesses were soon shown to be due to infection from air, water, food, animals or neighbours. This led eventually to the successful treatment or prevention of a great variety of scourges, from plague to tuberculosis. The heat treatment (pasteurisation) of milk to rid it of disease germs is familiar. Modern surgery is possible only because infection can usually be prevented. A historian has written that, when a Paris newspaper asked its readers who was the greatest Frenchman, 'the votes for the humble chemist outnumbered those for Napoleon, Charlemagne, and all the rest'.

The principle of Pasteur's experiments is easily understood and the experiments are not hard to describe. The reader (and the author) might, however, find them difficult to repeat: my account of them has to be misleadingly brief. A longer description would give a clearer picture, both of the experiments and also of the surrounding controversy; and the experience of growing, or preventing the growth, of microorganisms in the laboratory would lead to still better understanding. Most descriptions of scientific work in the following chapters have the same limitations. It is sometimes possible to outline the evidence for the conclusions put forward; and (often equally important) reasons can be given why some beliefs are rejected. But it is not possible to convey the slow, laborious procedures needed to achieve the results described.

The Origin of Living Things

Biologists allow one exception to the rule of all life from other life. Living things are assumed to have originated, on earth, from inorganic materials. Four thousand million or more years ago, it seems, the ancestor of all known organisms arose. The weather at the time was not what we should regard as favourable: it probably included continuous violent storms of hot rain, accompanied by lightning. Volcanic eruptions were numerous. The atmosphere contained little oxygen but plenty of hydrogen, ammonia, carbon monoxide and perhaps methane or marsh gas—for us, a more than toxic mixture which probably also stank of hydrogen sulphide. The ultraviolet radiation from the sun was not absorbed by a layer of ozone. If chemists could reproduce such conditions in a gigantic laboratory, they could perform some notable researches.

The main organic substances in cells

Small molecules
Amino acids (20)
Nucleotides (5)
Fatty acids
Sugars

Large molecules formed from smaller ones
Proteins (made up of amino acids)
Nucleic acids (made up of nucleotides)

Early Life

During hundreds of millions of years of storm, heat and radiation on the young earth's surface, we guess that small organic molecules were repeatedly formed. If methane (a carbon compound), ammonia and hydrogen are heated with water, and an electrical discharge (simulating lightning) is

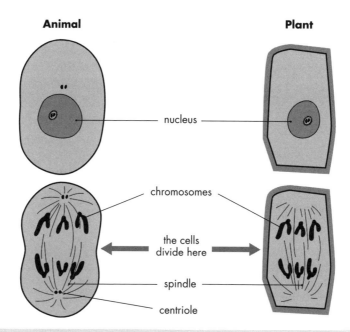

The core of modern biology. Cells of an animal (left) and a plant (right). In the 'resting' state (top) the only structure visible in a nucleus is a nucleolus. The deoxyribonucleic acid (DNA) is in chromosomes which appear when a nucleus divides. Division of the nucleus is usually followed by division of the cell. Nucleic acids, the physical basis of biological heredity, are present in all living things. This knowledge has solved many problems and has presented us with many new ones.

passed through the mixture, small organic molecules appear. Next, small molecules must have made larger ones. Clays can concentrate organic substances on their surfaces. If phosphates are present, they can promote the linking of small organic units, such as amino acids, in chains called polymers. Such processes require very unusual conditions but there was, presumably, plenty of time. The important substances formed by chains of amino acids are proteins—large, complex molecules found today only in organisms. Other molecules, also confined to organisms, are the nucleotides. The five present in all cells can, like amino acids, form chains and so produce ribonucleic acid (RNA) and deoxyribonucleic acid (DNA).

RNA and DNA can have their nucleotides arranged in various orders. DNA is the physical basis of heredity: the 'code' transferred from one generation to the next is the order of its nucleotides. In each cell, this information is transmitted to RNA. The RNA then specifies which of the twenty amino acids are included in a protein and the order in which they are arranged.

For the evolution of life, DNA molecules and associated proteins had to be surrounded with membranes consisting of oily substances, the phospholipids. In this way, it seems, cells were formed. Membranes are among the essential features of living things. At first, the RNA and DNA were evidently dispersed in each cell; bacteria remain in this state (called prokaryotic). The DNA in the cells of most organisms is inside a nucleus, which has a membrane of its own. This is the eukaryotic state. When a cell divides, the nucleus divides first. The DNA is then visible in structures, usually threadlike, the chromosomes, illustrated on page 9.

Conceivably, each major step in the origin of organisms occurred only once. If so, it was a 'quantum leap'—an improbable event which, because it is unrepeatable, is inaccessible to investigation. But why, the reader may ask, should these events not have occurred more than once? Perhaps they did. Why, indeed, should life not still be arising, perhaps in hot springs during thunderstorms? Unfortunately, newly created organic substances would soon become the food of existing organisms. The full series of steps needed to produce organisms from simple substances could not occur on earth today—except perhaps in the laboratory.

A feature of current speculation about early life is its strictly chemical basis. Until recently, it was quite usual to speak of a 'vital essence' or an *élan vital* (or some such thing) as inhabiting organisms and giving them life. The meanings of the Latin anima, from which we derive words such as animated, include health, life, person and soul. Ancient Greek had a similarly versatile term which, in our lettering, becomes psyche. But today, talk of essences or vital forces is not found useful. This is partly because all the chemical elements present in living things are also found in inanimate matter—in rocks, the sea or the air. Adding essences does not help our experiments or even our speculations.

Yet we do make a sharp distinction of the living from the nonliving. Living things are sometimes, even in scientific writings, likened to machines. This is misleading: machines are made by human beings to perform some operation, such as pumping, flying, weaving, weighing or telling the time. They can be constructed from printed diagrams. Their designers (and, sometimes, the people who use them) know how they work. They do not reproduce themselves: instead, if they are dismantled or damaged, they can be reassembled, perhaps with some new parts. Organisms, in contrast, are like Humpty Dumpty: they cannot be taken to pieces and put together again. Among other differences of an organism from a machine are the effects of use: exercise your muscles and other moving parts, and they do not wear out (as a bearing does in an engine) but become stronger; moderate friction on the skin does not scrape it away but stimulates extra growth.

However they originated, the organisms of today have in common most or all of the features shown in the table below.

Characteristics of living things

Chemistry: contain proteins, nucleic acids, fatty acids, sugars and water.
Metabolism and homeostasis: take in substances and pass others out while retaining the same or slightly changed bodily form.
Self-recognition: distinguish and reject foreign substances such as the proteins of disease organisms.
Sensitivity: 'feel', or respond to changes in their surroundings.
Reproduction: reproduce themselves.
Development: develop, often from a single cell, such as a fertilised egg, to something much more complex.
Heredity: transmit information from one generation to the next (not always quite accurately); hence offspring resemble parents.
Evolution: evolve—populations change with time and diversify.

Not all organisms have all these features.

The list may suggest that biology is a unified subject and, in some ways, so it is. But, for an opposite impression, one has only to go into a garden, a woodland, a rocky shore, a park or a field and look around. Organisms, though all composed of the same kinds of molecules, are immensely diverse. The number of species already described is said to be about two million. I have not counted the species myself but I accept this figure without a struggle. In later chapters we see something of this diversity; in the next, more on the still argued topic of how it arose.

CHAPTER 2

EVOLUTION: HISTORY OR BLASPHEMY?

Where order in variety we see
And where, though all things differ, all agree.

ALEXANDER POPE

SOME ENGLISH UNIVERSITY STUDENTS, well trained in conventional biology, were once challenged to defend the statement that organic evolution has occurred. All accepted evolution but, as it turned out, only as an axiom: none was able adequately to cope with counter arguments offered by two teachers acting as disbelievers. That was in 1950. A hundred years earlier, few undergraduates would have had any concept of evolution: most, perhaps, would have accepted that the universe is about 6000 years old; and that everything, including all species of organisms, had been created by a supreme being in a few days of intense activity. In 1654, a scholarly archbishop, James Ussher (1581–1656), had calculated the time of the creation, on evidence from ancient Jewish writings, as noon on 23 October 4004 BC.

The effect of Christian doctrine was reflected in attitudes to fossils. The petrified remains of animals had been known, and even explained, for millennia. Xenophanes of Colophon (about 560–478 BC), an early Greek philosopher, interpreted them as the vestiges of extinct forms. (Appropriately, Xenophanes was, in the words of the philosopher, Bertrand Russell, 'as regards the gods . . . a very emphatic free thinker'.) But when, in 1726, a giant fossil salamander was unearthed in Switzerland, the discoverer called it *Homo diluvii testis*, or 'man, evidence of the Flood'.

As we see shortly, similar notions are still with us. Why, nonetheless, do most of us accept evolution as something that has happened—the descent of living things in the distant past from other, very different beings?

HARD EVIDENCE AND LOST WORLDS

Our access to the past is through the study of the rocks which form the earth's crust. The rocks are layered. Rarely, the layers or strata can be seen without effort. Usually one has to dig to find them. In the simplest case, the deeper one digs, the older the rocks. A crucial observation on fossils was made in England late in the eighteenth century. Canals were being dug across the country and in England and Wales, conveniently, the strata are tilted. In the east, in Norfolk, the rocks at the surface are relatively recent (Pliocene); in the west they are ancient (Cambrian); those in the midlands are intermediate. A geologist and surveyor, William Smith (1769–1839), saw that each stratum has its own typical set of fossils and that these *always succeed one another in the same order*. As we examine rocks of increasing age, we find fossils less and less like modern, living forms.

The oldest fossils are of microscopic organisms which lived about 3500 million years ago and are found in stromatolites ('stony mattresses'). Ground to exceedingly thin, translucent sections and examined under the microscope, these rocks reveal the remains of several kinds of organisms, including bacteria. For 2000 million years only such simple prokaryotic organisms were present on earth. Cells with nuclei appeared about 1.4 thousand million years ago; many-celled organisms arrived several hundred million years later. After that, evolution speeded up.

Fossils are usually of hard parts, such as the fibres of plants, the shells of molluscs and the bones of vertebrates. When they are buried in mud or silt and covered by deposits of increasing depth, they may become petrified. A tissue such as bone, though tough, contains many spaces, often microscopic. These pores

A refutation of 'creationism': stratified rocks. These were deposited, in what is now Yellowstone National Park (Wyoming, USA), between about 60 and 40 million years ago. Fossils are found in a regular sequence in such strata; they give us an account of the evolution of life covering about 1000 million years.

are slowly filled with mineral deposits which give the fossil its hardness and weight but retain its shape.

The fossil record shows that, during the past 700 million years, many kinds of creature have flourished only to disappear. The earliest (Precambrian) rocks of this vast span contain traces of small creatures of which some seem to be jellyfish; others resemble worms. Then, at about 570 million years ago, the rocks are suddenly full of fossils. S.J. Gould, an American paleontologist celebrated as a populariser, in a dashing mixed metaphor writes of 'the first flowering of the Cambrian explosion'.

The many Cambrian fossils include crustaceans and other animals, some of which can be placed in familiar groups. Others cannot. When a picture of one of them, *Opabinia*, was first shown on a screen to a biological audience, it evoked astonished laughter. This creature, 70 millimetres long or less, has a trunk or nozzle and is segmented; each segment seems to carry a pair of gills; of the five eyes, four are on stalks. The gut is a straight tube but at the front it makes a U-turn so that the mouth faces back. One guess is that the nozzle picked up food and bent

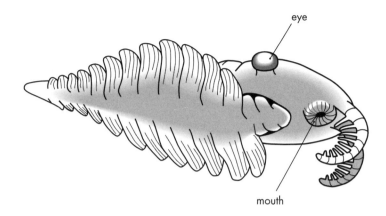

Extinct. A Cambrian weirdo from the Canadian Rockies: *Anomalocaris* was more than half a metre long and seems to have had a pair of stalked eyes; near the mouth were two appendages, presumably for feeding; the series of lobes perhaps carried gills and may have propelled the animal through the water. Although its name means 'unusual shrimp', *Anomalocaris* was not a shrimp or even an arthropod: it is given a phylum of its own.

round to the mouth. The largest known animal from that time, *Anomalocaris*, shown above, was about half a metre long, had an array of 'teeth' and was probably a predator.

Interpreting these ancient forms is hazardous. One, *Hallucigenia*, was believed to have had seven pairs of spines on which it walked and seven tentacles each with a mouth. But, in 1991, Chinese workers turned it upside down and showed it to be one of the velvet worms (Onychophora)—a small but well known phylum which has been thought, probably incorrectly, to represent a stage in insect evolution. The reader interested in this narrative should watch the scientific journals, especially *Nature* and *Palaeontology*, for the next thrilling instalments.

Let us move on. Three hundred and fifty million years ago, much of the land was densely covered with early land plants, the horsetails and club mosses; later, they were largely replaced by the tree ferns which, fossilised under pressure, now provide us with coal. In the coalmines we find fossil amphibians. Some, remarkably complete, were giants more than three metres long.

We now casually jump another enormous lapse of time: after 50 million years, reptiles evolved from amphibians. Reptiles are thoroughly fitted for life on land, even for deserts; and, on land and in the waters, they diversified. The most celebrated consequences are the dinosaurs, of which two major groups lived for about 200 million years.

Gigantic fossil dinosaurs began to be thoroughly studied early in the nineteenth century. *Iguanodon* was reconstructed (not quite correctly) in London in 1853 and, to celebrate, a dinner party of 21 persons was held inside it. It was a biped and, at a height of over 4 metres, it must have been more than impressive. The nearest in appearance we have today are monitors of the genus, *Varanus:* to see one of these carnivorous lizards, the height of a man, moving swiftly on its hind legs, is exciting enough.

Some dinosaurs were only the size of a chicken. At the other extreme were the notorious monsters. Until 1995, the largest carnivorous dinosaur known was *Tyrannosaurus* of about 7 tonnes—ten times the weight of a male polar bear (*Thalarctos maritimus*) which is the largest existing flesh eater on land. Now, from the Cretaceous of Argentina, we have *Giganotosaurus*, probably 8 tonnes and more than 12 metres long. How did such a massive creature feed? Perhaps mainly by scavenging. More massive still were the plant eaters such as *Brachiosaurus*—25 metres long and weighing at least 50 tonnes or ten times the weight of an African elephant; at 13 metres it was twice the height of the tallest giraffe.

The giants have left many footprints in mud which was later petrified. Their gait resembled that of a mammal or bird, not that of a crocodile with the legs splayed out. The largest probably moved at the reader's slow walk (say, 1 metre a second); the rather smaller *Apatosaurus* (or *Bronchosaurus*) may have achieved a slow run, like that of an elephant. The 'ferocious' bipeds, weighing a mere half tonne, perhaps managed 11 metres a second. (Compare a race horse's 16 m/sec.) R. McNeill Alexander, a Scottish zoologist who has made technical studies of these creatures, believes that he could outrun a tyrannosaur. Alas, despite DNA, little hope exists of putting him to the test.

Dinosaur behaviour and physiology are enigmatic. Most reptiles today are coldblooded (poikilothermous or ectothermic). To become active after a cold night, they need to warm up: lizards often bask in the sun. But reptiles do not sweat and can overheat: on a hot day they may choose shade. The large dinosaurs probably had raised internal temperatures. *Stegosaurus* has massive plates on its back which were once called armour— an instance of the common belief in nature as implacably

Armour, or a cooling device? *Stegosaurus*, at 6 metres one of the longest dinosaurs, had massive plates on its back. What were they for?

violent. Today, partly as a result of experiments with models, they are believed to be a means of losing heat. If so, how did other giant reptiles manage without them?

Adaptive radiation took some reptiles back to the water. The many kinds of longnecked, longtailed plesiosaurs developed flippers but remained obviously reptilian. Some plesiosaurs were 13 metres long. A persistent story suggests the survival of monsters of this sort in large lakes, such as Loch Ness in Scotland, but 'Nessie' remains a myth (or a fraud). Other marine reptiles, the ichthyosaurs, went back to looking just like fish.

On land, among the many weird sights of the age of reptiles were the airborne pterodactyls (order Pterosauria). They ranged from sparrow-sized species to *Qetzalcoatlus* with an estimated wing span of 11 metres—similar to that of an executive jet. Another, *Pteranodon*, had about the same wing span, a tremendous, elongated head and very thin and light bones. It probably weighed only 17 kilograms. These giant forms, perhaps warmblooded, may have kept themselves airborne by soaring on thermal currents, as some large birds do today.

From a lost world. The skeleton of a pterosaur, *Nyctosaurus gracilis*, from Cretaceous rocks. Pterosaurs were mechanical marvels and could certainly soar as an albatross can; but it is not clear how they took off: perhaps from cliffs.

More than 60 million years ago, dinosaurs, pterodactyls and many others disappear from the fossil record. The apparently abrupt extinction of all these lineages has encouraged speculation. One catastrophic theory is global destruction due to the impact of bodies, such as asteroids (bolides), from space. A change of climate has been proposed as an alternative; but just what change, and how it operated, is unclear. I am sorry: I do not know how the dinosaurs became extinct.

The many fossil sequences known can be given an acceptable explanation only if we assume evolution. They also tell us something about the patterns of evolutionary change. A straight line of seeming improvement, taken by itself, is usually misleading. The horses are an example. Some textbooks give the impression that one member of the horse family is present in each geological epoch; each successive type then represents a step toward the 'perfect' modern horse. In fact, however, at almost any time during the past 60 million years we find many horsy forms. A complete chart of the horse family looks like a tree with many branches. Each major lineage shows similar changes: enlargement of the body; loss of digits on fore and hind limbs; progressive adaptation of the teeth for grinding grasses. Every branch is cut off before the present, except for the one bearing the modern horse, the ass, zebras and the now extinct quagga, all of which come in the genus, *Equus*.

THE EXCITEMENTS OF CLASSIFYING

Organic evolution, with its vast time scale, came to be widely accepted only in the second half of the nineteenth century. Before then, collectors and classifiers had already assembled the living world in species, genera, families, orders and so on, as in a family tree. In the seventeenth and eighteenth centuries they were encouraged by the increasing numbers of strange plants and animals brought by explorers to European universities and museums. Some were of economic value.

Of the many classifiers, the most famous was a Swede, Carolus Linnaeus (1707–1778), a physician originally intended for the ministry. After struggling with poverty he became a professor of botany and arranged for many young naturalists to venture into distant regions, including Siberia, China and Japan. Some died of tropical diseases, but much material was collected. Linnaeus himself travelled in Lapland, north of the Arctic Circle, where he found and described many strange plants. He also studied local customs, especially the use of plants for culinary purposes. One was an insectivorous plant (*Pinguicula*) of which the leaves were used to curdle milk. He was therefore a founder of ethnobotany—a discipline now crucial for our attempts to preserve the living world.

Linnaeus was a complex character. His classification of plants is an orgy of eroticism. He writes of the 'betrothal' of plants and likens their reproductive organs to those of human beings. The calyx of a flower is equated with a woman's labia majora or a man's foreskin; the petals become the labia minora. Before marriage, trees and shrubs put on wedding gowns. Flower petals make bridal beds. Yet in private life he was straitlaced. He forbade his daughter to attend court, for fear of exposure to immorality. Nor was she even permitted to go to school; perhaps he thought that she would be taught botany.

Immense industry and personal charm enabled Linnaeus to persuade the learned world to name each species with two Latin words, such as *Canis familiaris* for the domestic dog. (The decline of classics and the discovery of many new species have obliged modern systematists to extemporise, with some odd results. In 1980 a new species was named *Onyx barnetti*. It is, I am told, a small worm inhabiting mangrove swamps in Queensland.)

In the Linnean system, the second term indicates the species, the first, the genus. The canine genus includes the wolf, *Canis lupus*, and others. Genera are grouped in families. In the family of dogs are also *Lycaon* (which includes the African Cape hunting dog, *L. pictus*) and the South American *Speothos*. Dogs (Canidae), cats (Felidae), bears (Ursidae) and others make the order, Carnivora. Such a system fits the idea of evolution.

The Linnean system may appear cumbrous but the Latin names it replaced were worse, for they often attempted a full description. A beautiful tree, called in English the red maple, was named *Acer Americanum, folio majore, suptus argenteo, supre viridi splendente, oribus multis coccineus*; today, it is *Acer rubrum*. Nonetheless, an English zoologist who insists on referring to a robin as *Erithacus rubecula* may seem absurdly pedantic. But suppose her audience includes visitors from Italy: they will be accustomed to calling a robin, 'pettirosso'; or, from Germany, 'Rotkehlchen'. Worse, in the United States a robin is not *E. rubecula* but a thrush with a red breast, *Turdus migratorius*.

The last example is one of many resulting from the emigrations of English-speaking people. Early British visitors to India found many strange birds. One has a black body and brilliant chestnut wings—a disconcerting sight when it turns up in a patch of jungle. In Hindi it already had two names, 'mahoka' and 'kuka'. The new arrivals called it a crow-pheasant. Its internal structure, however, reveals it as neither a crow nor a pheasant, but a cuckoo (family Cuculidae); its international name is *Centropus sinensis*.

The naming of plants has not been easier. Common names have varied, sometimes for reasons of decorum: a lily, naked ladies, was disguised as 'lords and ladies' before it became *Arum maculatum*. Latin names such as this sometimes aroused resentment. An eighteenth century versifier complained:

> The flaunting woodbine revell'd there
> Sacred to the goats; and bore their name
> 'Till botanists of modern fame
> New-fangled titles chose to give
> To almost all the plants that live.

Goat's beard had become *Tragopogon pratensis*.

The system of biological naming has now overcome all

objections and is a notable example of global cooperation. It is essential not only for academic use but also for medical and economic biology. The microscopic organisms which cause malaria are of several species. The infection is carried by many species of mosquitos. To manage or to prevent the disease, one must know exactly what species are present. The same applies if one is trying to encourage the growth of forests; or to increase the numbers of edible fish; or to prevent attacks by pests.

THE PAST IN THE PRESENT: STRUCTURES AND EMBRYOS

Many strange features of organisms would oblige us to assume evolution, even if we had no fossils. A human designer would not use the same set of components for a paddle for swimming, a limb for running and a wing for flying. The limbs of birds and mammals include hands, flippers, wings and the powerful columnar legs of elephants, yet their skeletons are all variations on the same pattern: five digits (fingers or toes) on each limb, supported by a set of bones remarkably similar to those of our own limbs. We attribute this strange correspondence to a common ancestor, with pentadactyl limbs, for all vertebrates that live on land. Nor would an engineer give every mammal the same number of neck vertebrae. Yet most mammals have seven: those of a whale, which has virtually no neck, are flat plates; those of a giraffe are greatly elongated; human cervical vertebrae come in between. Again we assume a common ancestor, with seven neck vertebrae, for all mammals.

The bones of a vertebrate's limbs and neck illustrate homology. The lower part of the forelimb of a horse is the homolog of the human middle finger. The earliest horsy mammals had five fingers and toes; more recent fossil predecessors of modern horses had three; today, apart from vestiges, only the middle one remains. Embryonic development tells a corresponding story. Each limb of an embryo horse begins with five rudimentary digits—an example of the recapitulation of ancestral features during development. All except the middle finger and toe almost completely disappear long before birth of the foal.

At one time all the stages of an embryo were believed to correspond exactly to the stages of evolution. They do not; but

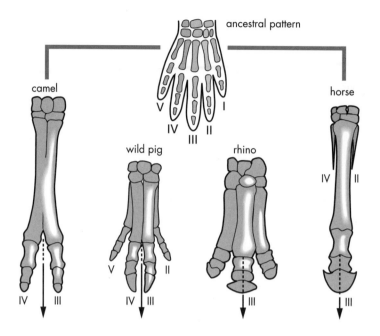

The five-digit or pentadactyl limb occurs in all land vertebrates. (But in some, such as snakes, it is seen only in the embryo.) The top figure shows the fundamental plan of a hand or foot. The other figures are of forelimb skeletons: from left, camel, pig, rhinoceros, horse. All, like our own, are variations on a single theme.

many others do reflect the evolutionary past. The vertebrates fully adapted to life on land, the reptiles, birds and mammals, all have closely similar early embryos: they have pouches which look as though they will become gills; blocks of rudimentary muscles which could form the segmented musculature of a fish; a muscular tail, again like that of a fish; and—another fishy feature—a simple heart which pumps blood forward. So the embryos at this stage all look as though they are about to become something like a salmon or a cod. But a structural reorganisation quickly follows and the adult form begins to appear. Correspondingly, the fossil record shows land vertebrates evolving from fish.

Stranger findings have come from recent experiments. The fossil record indicates a lapse of tens of millions of years since the ancestors of birds had teeth, yet the cells of modern birds can still produce them. Tissues from a bird's embryo beak have been grown with cells from the mouth of an embryo mouse.

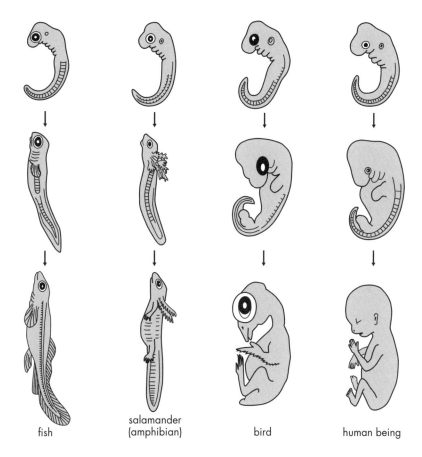

fish   salamander (amphibian)   bird   human being

The embryos of animals often reflect their evolutionary past. This famous picture, drawn by the German biologist, Ernst Haeckel (1834–1919), shows the similarity of all early vertebrate embryos (top row). The resemblance is explained by descent from a common ancestor.

Under this stimulation, teeth began to grow from the bird's tissues. None of this makes sense unless we assume evolution.

A Struggle for Revival? Myths of Creation

Despite all the evidence, an energetic movement, active especially in the United States, asserts that evolutionary biology is both factually and morally wrong. The two creation myths in *Genesis*, the first book of the Bible, are represented as the true account of how *Homo sapiens* and other organisms came into existence. Some creationists insist that evolutionary biology is heresy and should not be taught at all. The teaching of evolution, they say, is a source also of anarchy, militarism, communism, nazism, racism and other ills. Alternatively, some American schools are required to give equal time to the biblical stories and biology.

Genesis, however, is an alternative not to biology but to other myths. Greek writings of the classical period offer several. In one, Mother Earth emerged from Chaos and bore a son, Uranus. He showered his mother with rain, and she bore trees, grass and flowers and the beasts that go with them. Hinduism similarly has male and female founders: Vishnu, the male, is a protector; Lakshmi, his consort, is a goddess of abundance. From Vishnu emerges Brahma, each of whose four arms holds one of the sacred books, the Vedas. Other beautiful and awe-inspiring stories of creation exist, handed down in speech or in ancient writings. They represent early attempts to find and to explain order in the confusion of our surroundings.

Biological accounts of evolution, too, try to show order among living things. They arise, however, not from sacred texts but from the systematic study of fossils and of existing organisms. Although biology is sometimes taught dogmatically, like all science it is based on findings which are regularly debated and questioned. Heresy is not a component of the scientific method.

The creationists are in effect opposed not only to evolutionism but to science. They speak of 'creation science' yet they hardly ever submit findings from their studies to scientific journals. Instead, they appear on television and publish extensively in the news media. Much of their audience is without the knowledge to refute them, hence creationists are free to make false statements on scientific questions. Some say that no intermediate forms have been found in the fossil record—an absurd untruth.

Creationists also cite biologists' disagreements on the *rate* of evolutionary change as evidence against the *occurrence* of evolution. They should be reminded of the commandment, in *Exodus* 20:16, against bearing false witness.

AN INTERMEDIATE FORM

Archaeopteryx, the earliest known fossil with feathers, was at first classified as a reptile. It was about the size of a jackdaw (*Corvus monedula*). It had solid bones, teeth and a long bony tail—characteristics of reptiles, not of modern birds. It resembles lightly built, bipedal and seemingly agile carnivorous dinosaurs called coelurosaurs. It is usually believed to have clambered and glided, flapped or flopped about in forests. (The fossilised individuals seem to have been blown out to sea.) One paleontologist, J.H. Ostrom, however, has suggested that birds' wings did not originate as a means of gliding: on his view the earliest feathered reptiles were small dinosaurs which ran about on their hindlegs and used their forelimbs to catch insects. But the idea that wings originated as butterfly nets, though well argued, is not universally accepted.

Archaeopteryx is a link, not missing, between ground living reptiles and the superb airborne creatures we know today. Its first discovery came opportunely two years after publication of Darwin's *On the Origin of Species*. T.H. Huxley (1825–1895), Darwin's most prominent supporter, had just stated that such an intermediate form must have existed.

Faced with difficulties, some creationists resort to the bizarre. Fossils have been explained as devices provided by the devil to deceive us. They have also been said to be challenges by the Almighty to test the faith of believers. Sometimes creationists seem to parody their own writings. The sequence of fossils in successive layers of rock has been attributed to the flood described in chapter 7 of *Genesis*. It is hardly surprising that many Christians regard creationists as blasphemers.

Nonetheless, creationists should not be brushed off as merely a tiny, eccentric minority. An exceptionally popular United States president, Ronald Reagan, once pronounced in their favour. More important, American biological texts are still influenced by Christian fundamentalism. Few teachers of biology in American schools are familiar with even the elements of evolutionary theory. As a result, in some States of the USA, surveys have revealed more than half of all undergraduate students to be creationists. The remedy is education in logic, in the methods of science, in biology and in moral principles.

✻

The fossil record shows organisms spreading over the world and becoming more diverse. The results seen today are species marvellously adapted to a great range of situations. One may therefore accept the reality of organic evolution but still ask what has impelled these vast changes. The reader may also complain that I have barely mentioned the principal founder of evolutionary biology, Charles Darwin (1809–1882). So here is a brief statement on *how* organisms have come to be what they are (more in chapter 10).

Individual variation is found in every species. Among animals, the most familiar examples are the varieties of domestic cats, dogs, cattle, sheep, camels and the rest; and, among the plants, the crops in the fields and the fruits and flowers in our gardens. Stockbreeders, farmers and gardeners select from these types for breeding. Often, the offspring of a particular form resemble their parents. Breeders can then, over generations, change the selected forms in a planned direction. Evolution is held to be due to a similar process, but not planned and much more prolonged. Ever since living things began, slow changes have evidently been responsible for the millions of species which now exist or have existed. Darwin called the process natural selection. This was his fundamental contribution to biological theory.

It therefore makes sense to ask what is the *function* of any feature of an organism: that is, what is its contribution to the survival or breeding of the individuals that possess it? This is a prominent difference of living things from nonliving. We do not regard the hardness of diamond or the heat of the sun as having functions in this sense.

During much of the past 4000 million years, only microorganisms have existed, some green, some not. But, eventually, through the working—it seems—of natural selection, these gave rise to the variety we know today: green scums and giant redwoods; microbes and moulds; amebae, ants and apes. The existing species are fitted to environments which range from cracks in hot rocks to deserts of ice and from the abyss to the tops of mountains. But, despite their diversity, still incompletely known, all have one structural feature in common: cells.

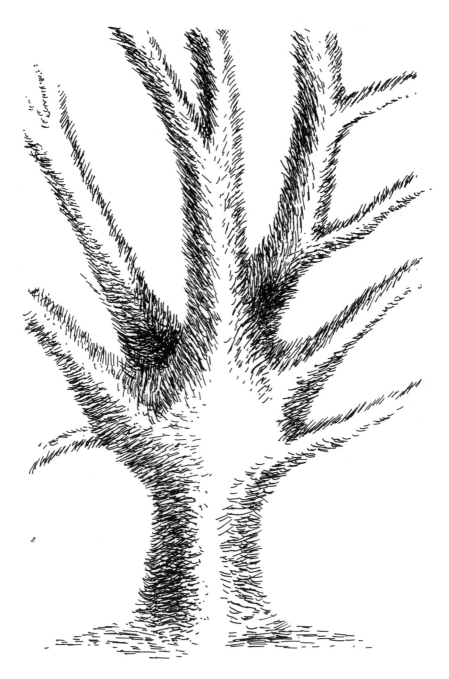

'Engineering' the genes. Alas, I have not yet succeeded in creating a fur tree, *Arbor villosissimus*. If I had, I could have sold the patent for millions and endowed longterm ecological research.

THE UNSEEN WORLD

In the rest of this book we begin with the smallest living phenomena and, with some digressions, work up to large ones.

What we can see of the inside of a single cell is as complex and enigmatic as the constellations of a night sky. Yet much is now known about how cells work and how they transmit information from one generation to the next. One result is a great increase in knowledge of how the body resists infectious diseases and of cancer. Another is the ability to make drastic improvements in the organisms we eat. There is also the possibility of creating new plants and animals.

Meanwhile, longstanding questions remain: one is how heredity and environment interact; another is how to cope with the unseen organisms which surround us and on which we depend.

CHAPTER 3

CELLS, SEX AND DNA

Small is beautiful.

E.F. SCHUMACHER

IN SEVENTEENTH CENTURY EUROPE, the English puritan, John Bunyan (1628–1688), called consumption 'captain of all the men of death'. Today, we call it pulmonary tuberculosis. The very small tubercle bacillus was discovered by the German bacteriologist, Robert Koch (1843–1910), in 1882. He devised a new method of colouring microorganisms so that they could be clearly seen; he also benefited from the new microscopes of Carl Zeiss (1816–1888), founder of the famous optical works in Jena. Koch's work, which led eventually to the near conquest of tuberculosis in rich countries, is an example from the many advances in biology and medicine to which we return later. It depended on seminal studies of the treatment of cells to make them visible and of cell chemistry.

The most notable of the early microscopists was a Netherlander, Anton van Leeuwenhoek (1632–1723). He was destined for trade, but instead he accepted a lowly post from the municipality of Delft. Most of his long life was dedicated to making microscopes, for which he ground the lenses, and to describing what he saw with them. He examined water from the sea, from ponds and from gutters; the tissues of animals; the scrapings from his teeth; and much else. Since he was able to magnify objects up to 270 times he saw not only protozoans, such as amebae, but also bacteria. He described red blood cells and the fine capillaries through which they pass on the way from arteries to veins; he and an associate identified spermatozoa; and he made original observations on plant tissues. He was the outstanding naturalist of the unseen living world.

Microscopy and Ultramicroscopy

The universal cell structure of animal and plant tissues was, however, established only in the nineteenth century. The leading figures were two Central Europeans who looked closely at the insides of cells, not just their walls or outer membranes. M.J. Schleiden (1804–1881), son of a physician in northern Germany, qualified in law but failed as an advocate. He became severely depressed and shot himself in the head but survived. He then took up science and, despite persistent melancholia, became a professor of botany. His main contributions concerned the cell nucleus and the development of the embryo of flowering plants.

Theodor Schwann (1810–1882), was born in Prussia but held chairs in France and Belgium. Unlike his friend, Schleiden, he was a cheerful man. He disliked controversy and refused posts in German universities because, he said, German microscopists were too quarrelsome. His system was based on that of Schleiden. He discovered embryonic cells in tadpoles; these, he found, reproduced themselves by division in much the same way as Schleiden's plant cells. He also saw that many tissues are cellular although they do not appear so. An example is muscle, in which the cells are greatly elongated fibres. These two men were therefore principal founders of histology—the microscopic study of tissues.

What we see of cells under the microscope is complex, beautiful, often baffling, but invaluable. In many kinds of illness, the cells and tissues change in typical ways. Hence histology soon came into medical use. The tissue most used in diagnosis is blood, illustrated on page 34. The level of hemoglobin and the number of red cells in each cubic centimetre of your blood may be below normal: if so, you are anemic and the cause should be looked for. An abnormally high red count is also possible and is advantageous if you live at high altitude, for

Making Cells Visible

Animal cells are typically ten to twenty thousandths of a millimetre in diameter—about one-fifth the size of the smallest objects visible to the naked eye. They are also colourless and translucent and are barely visible even if much magnified. To study them under the microscope, a small fragment of tissue is usually first dried and fixed in a preserving fluid—preferably one which does not cause severe shrinkage or distortion. The fragment is next embedded in a wax-like substance or a resin. The resulting firm block is cut into very thin sections. These are mounted on slips of glass ('slides') and the sections are stained with dyes. Some of the internal structures of cells can often be stained in contrasting colours.

instance, in the Andes, where it enables you to cope with shortage of oxygen; otherwise it may be due to excessive production of red cells in blood forming tissues—a rare condition (polycythaemia vera), cause uncertain. More important, disorders of blood formation can produce excessive numbers of white cells (leucocytes). This is likely to be one of the leukemias—now treatable by an impressive array of chemical agents. But an excess of one kind of white cells (eosinophils) suggests an allergic disorder. Among many other possibilities, parasites, such as those of malaria, may be present. In a clever crime story set early in this century, the detective, who is also a physician, identifies the microfilariae of elephantiasis in blood at the scene of a crime. The presence of these minute worms enables him to describe the criminal.

Traditional microscopy remains essential; but today the electron microscope shows us the insides of cells in much more detail. It uses beams of electrons instead of light and gives us photographs of cells at far greater magnification than before. As a result, the cell's internal components have been revealed as the complex structures shown on page 35.

The Plant Cell: Drugs and Flavours

Some of the earliest studies of cells were on plant tissues. In 1663, Robert Hooke (1635–1703), an English physicist and engineer, saw cork under the microscope. (Cork cells have

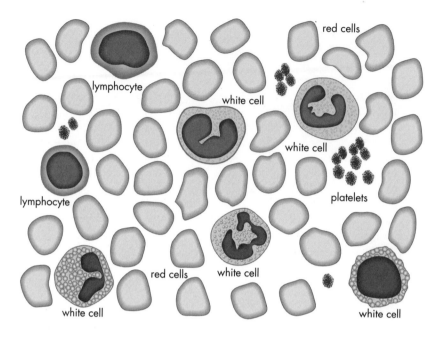

The fluid tissue: some blood cells. A cubic millimetre of normal adult human blood contains about 5 million red cells, 7500 white cells and 250 000 platelets (fragments of cells). Among the clues to a person's state of health are the numbers of red cells in a given amount of blood; the amount of hemoglobin in each red cell; the numbers of the many kinds of white cells; the presence of abnormal cells; and the numbers of platelets.

unusually thick cell walls.) Two years later, in his famous work, *Micrographia*, Hooke described observations made with early compound microscopes. His subjects included seeds and insects as well as cells. Not all academics welcomed his findings. One, a priest and the Oxford University Orator, attacked 'the new philosophy': the scientists, he said, could 'admire nothing except fleas, lice and themselves'.

The walls of plant cells are mainly cellulose, the carbohydrate which makes the main constituent of paper. The chemically active part of the cell consists of the cytoplasm and a nucleus. In the cytoplasm are structures (organelles) in which chemical processing takes place. Most fully developed plant cells also have large vacuoles filled with fluid. In rubber trees (*Hevea brasiliensis*) the vacuoles contain rubber. Other species offer poisons or strongly tasting substances which help to protect the plants from animals. Members of the family of cabbages and

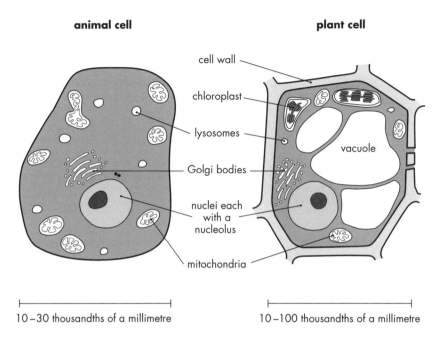

Modern knowledge. Some of the structures of cells, revealed by the electron microscope. Each drawing represents a thin slice through a cell. Every internal structure is a site of intense activity. On left, a typical animal cell. On right, a plant cell with its rigid cell wall and chloroplasts containing chlorophyll; it also has a large vacuole crossed by strands of cytoplasm. The vacuole may contain exciting substances, such as stimulants, tranquillisers, flavours or poisons. Molecules undergoing change move incessantly about the cell by diffusion.

radishes (Brassicaceae) contain oils (glycosides) which give us pungent flavours such as those of mustard and horseradish.

Plant poisons are important in medicine. Toxic glycosides, extractable from a handsome wildflower, the foxglove (*Digitalis purpurea*), have long been valued as a treatment for heart disease. As well, alkaloids, such as opium, caffeine and nicotine, are stored in vacuoles. Opium comes from the sap of a species of poppy, *Papaver somniferum*, known only in cultivation. Among its alkaloids are morphine and codeine. These pain-killers can also assist sleep; but, taken regularly, they may induce crippling dependence.

Caffeine, in contrast, in coffee or tea, if consumed with restraint is a valuable stimulant. It is also present in chocolate. If I now write of the chocolate tree, I am not digressing into a fairytale. The cacao tree (*Theobroma cacao*) is a native of Central

American forests. *Theobroma* means food of the gods. The Aztecs drank cacao as a bitter stimulant; and, after Hernán Cortés (1485–1547) had brought seeds back to Spain, so did Europeans. (Coffee came later.) The sweet drink was invented only in the eighteenth century and solid chocolate late in the nineteenth. Since then, cacao has made a worldwide industry. Yet much of its biology long remained obscure. Late in the twentieth century, entomologists have revealed its dependence for pollination on barely visible midges (*Forcipomyia*).

Among other seductive substances is nicotine. In the pure state, derived from tobacco plants (*Nicotiana tabacum* and others), it is exceedingly poisonous. In tobacco smoke, absorbed through the lungs with other substances, it kills millions and injures the health of many more. Another such poison, hemlock, from a herb, *Conium maculatum*, is celebrated for its use in the execution of Socrates (469–399 BC) by the authorities in Athens. (Socrates' offence was asking morally and politically awkward questions.) Hemlock causes paralysis by inhibiting impulses from the nerves to the muscles.

Other nerve poisons have several uses. Curare, from shrubs of the genus, *Chondrodendron*, was put on the tips of arrows or darts in South America. It can also be used to relax a patient's muscles during an operation. The reader, however, is more likely to have been dosed with atropine, from *Atropa belladonna*, the deadly nightshade: dropped into the eye, it paralyses the pupillary muscle, dilates the pupil and makes it easy for a physician to inspect the retina. The effect wears off in a few hours.

Or, if you have been in a malarial region, you may have dosed yourself with quinine, an alkaloid from the bark of one of the forty species of *Cinchona*, a South American genus. When bark from Peru was brought to England in 1654, Protestants objected, because it was a Catholic method. After French chemists had isolated the pure alkaloid in 1820, it was successfully used for many decades.

ENZYMES

The new knowledge of cells combines study of structure with chemistry. The chemical activity of cells depends on enzymes.

A crucial finding was made in 1897 by Eduard Buchner (1860–1917), a professor of chemistry in Berlin. He was making an extract of yeasts for medical use. To preserve it, he added sugar. The mixture was completely free of yeast cells, yet it fermented: alcohol was produced. Previously, the complex chemical changes in cells (such as alcohol synthesis) had been attributed to obscure vital forces. Now it became possible to identify the substances which produced them.

Enzymes are present in all cells and also in secretions such as saliva and those in the stomach and intestines. Most are proteins. They are catalysts and act by speeding up chemical changes while they end up unchanged. Each has a single action; its molecule precisely fits the substance (substrate) on which it acts. In the conventional metaphor, the relationship resembles that of lock and key.

Inside the cell, an enzyme usually takes part in a chain of reactions. Such chains can be summed up in simple equations such as those below. The first stands for cellular respiration—the process, almost universal in organisms, which yields energy; the other shows what happens when chlorophyll promotes the synthesis of sugars in a leaf. Each depends on the action of many enzymes.

$$\text{oxygen} + \text{sugars} \rightarrow \text{carbon dioxide} + \text{water} + \text{energy}$$
$$or \quad 6O_2 + C_6H_{12}O_6 \rightarrow 6CO_2 + 6H_2O + \text{energy}$$

$$\text{carbon dioxide} + \text{water} + \text{energy} \rightarrow \text{sugars} + \text{oxygen}$$
$$or \quad 6CO_2 + 6H_2O + \text{energy} \rightarrow C_6H_{12}O_6 + 6O_2$$

Organisms maintain their distinctive form, by self-correction, in spite of the continuous changes which go on inside them. So do cells. Stability depends on negative feedback in the many metabolic pathways. The principle is familiar from a thermostat switching off and on as its temperature rises or falls. Failure can lead to destruction of the system (as when a boiler overheats and blows up). In a cell, metabolic processes are precisely controlled. Negative feedbacks ensure that each process stops when the product exceeds a certain level. Unlike a thermostat, they operate with virtually no delay.

Negative feedbacks also enable cells to adjust their metabolism to changed circumstances. And, if one fails, often another can come to the rescue. That is how we survive 'the thousand natural shocks that flesh is heir to'.

Negative feedbacks operate continually. Most of the time the reader's cells are breaking down glucose derived from food. The liver also has a reserve, in the form of glycogen ('animal starch'). During violent exercise, liver cells produce extra glucose which goes in the blood to the muscles. One accompanying change is an increase in the number of mitochondria (shown on page 35) in the muscle cells. An adult has about enough glycogen for one day's typical exertions. In a fast, for a longer period, when no glucose reaches the cells from food or the liver, glucose is formed from fats or from amino acids derived from proteins. (Most of us have enough fats to last for several weeks.) Hence, in starvation, the fatty tissues of the body waste away and, later the muscles. But the integrity of the body and of most of its cells is maintained; and, when food is again eaten, the body can return to its usual state.

Enzymes are now put to a wide range of uses outside cells. On the farm, they are used to break down indigestible components of animal feed. This reduces costs by allowing the use of cheap food grains. In the laundry, some detergents include enzymes which break down fats (lipases) and proteins (proteases) and so deal effectively with greasy stains.

Self and Not Self: Immunity

For self-maintenance, cells must recognise and counteract invading foreign proteins and other large molecules. Suppose the reader is bitten by a king cobra (*Naja hannah*) of southern Asia. (This, at nearly 6 metres, is the longest venomous snake.) The venom contains a poison which prevents conduction of nerve impulses to the muscles. The bite is followed by paralysis of the heart and other muscles and death in about two hours. In this case, the body's resistance is overwhelmed by the massive dose of poison. The toxic substance is an antigen.

Death can be prevented by immediate injection of antivenin, produced by first inducing a cobra to discharge venom into a glass vessel. This skilled procedure, impressive to watch, is regularly carried out before an audience in the Haffkine Institute in Bombay. Small, harmless amounts of the venom are next injected

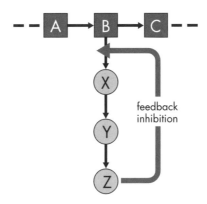

Automatic regulation by negative feedback in the cell. X represents the first enzyme in the pathway which leads only to the product Z. When Z reaches a certain concentration, the synthesis of X is stopped; Z therefore does not accumulate beyond a fixed level.

into a horse. Antibodies are formed in the horse's blood and a solution with the antibodies can then be extracted from the blood and stored. Injected into a patient, the antivenin combines with the antigens in the venom and makes them harmless.

The production of antibodies in a horse's blood is an immune response. In this case no organisms enter with the venom. More important is the response to disease organisms. They may be bacteria, protozoans or larger parasites. All consist of, or contain, antigenic molecules. If the infected person or animal is to survive, the pathogens must be killed or otherwise put out of action. Very rarely, an infant is born unable to produce antibodies (severe combined immunodeficiency or SCID); in the absence of elaborate treatment, the child becomes heavily infected and usually dies within a year.

When, in the nineteenth century, the importance of microscopic disease organisms was realised, research was concentrated on bacteria, which can be seen with an ordinary compound microscope. Then, late in the century, during study of a disease of tobacco plants, a crucial discovery was made: the infection could be transmitted by a liquid freed of bacteria by being passed through a fine filter. Soon, a similar finding was made on foot and mouth disease of cattle. As a result, the existence of pathogenic ultramicroscopic particles, now called viruses, was deduced.

Viruses are complicated fragments of organic matter, most of which can be seen (that is, photographed) only with an electron microscope. Most kinds have an outer layer of protein and fatty substances. Inside is either DNA or RNA. Like bacteria, they are present wherever there is life but, unlike bacteria, they are not independent organisms: they can multiply only inside a living cell. Those that cause disease usually destroy the cells they enter. Some, the bacteriophages, infect and destroy bacteria. When an infected cell collapses (undergoes lysis), it releases the viruses; they can then move on to other cells. This is what happens in those common afflictions, cold sores, caused by the herpes simplex virus. The same applies to a condition which until recently was grimly familiar: the skin eruptions of the deadly infection, smallpox, arise from destruction of cells.

Cellular Warfare

The most widespread kind of immune response among animals is by phagocytes, cells which engulf and digest intruding micro-organisms and viruses. Their importance was first shown clearly in 1882, by a Russian zoologist, Ilya Metchnikoff (1845–1916). While in Sicily, he was doing 'useless' academic research on the larvae of starfish. The larvae are transparent and contain mobile cells which resemble amebae (page 90). Metchnikoff thought that such cells could act as a defense against foreign objects. He inserted fragments of wood into larvae and the next day the splinters were, as he had predicted, surrounded by ameboid cells. This began a quarter century of fruitful study of phagocytes, mainly in the Pasteur Institute in Paris.

Later, the immune response of vertebrates was shown to depend primarily on different white blood cells, the lymphocytes. The blood of a human being contains so many that, put together, they would make a mass like that of the brain or the liver. They develop in the bone marrow and the thymus; these are the primary lymphoid organs. Lymphocytes are present not only in the blood vessels but also in the lymphatics—vessels which contain a fluid (lymph) but no red blood cells. They are carried in the lymph to the spleen, lymph nodes and other secondary lymphoid organs. There they are ready to react with

Pestilence: HIV Infection and Others

Some viruses exist in many forms and change rapidly. A vaccine for one form is ineffective for another. The virus of the common cold has more than a hundred varieties, which is one reason why, late in the twentieth century, we go on sneezing. Influenza remains dangerous for the same reason. Unlike colds, influenza is a killer, especially of the aged: in 1919, as an appendix to the first world war, it killed 20 million people—more than died in the war. Vaccines confer immunity for at best only a year.

Today the most threatening virus infection is acquired immunodeficency syndrome, or AIDS, now the leading sexually transmitted disease. A new and alarming illness was first suspected in 1979. The virus was identified in 1983, at the Pasteur Institute in Paris, and later named human immunodeficiency virus, or HIV. Its main targets in the human body are the helper T cells and some other cells of the immune system. As a result, an infected person succumbs to many diseases which are usually resisted. A delay of about ten years commonly occurs between infection and the onset of fatal illness. This seems to be a result of an at first successful immune response. But people vary greatly: some show effects of the virus much earlier. Others among those at risk seem able to resist the infection indefinitely. Study of these fortunate people may help to show how to promote resistance.

Some drugs delay the progress of the disease but, at the time of writing, no effective vaccine has been developed. A major difficulty is the high mutation rate of the virus: each infected person may have a different form.

Perhaps the major technical problems will soon be solved. But HIV infection is not merely a matter of cell physiology. We can see this if we ask how the disease arose, how it is transmitted and who are the most numerous sufferers. In 1973, before AIDS had been heard of, an African anthropologist, A. Kashamura, published an account of the customs of his countrymen near the great African lakes. One popular ritual was to inject a man or woman with the blood of a monkey of the same sex: this was supposed to induce vigorous sexual activity.

Since then, viruses resembling HIV have been identified in African monkeys and apes. Whether they were the origin of AIDS is not known. The hypothesis of transmission from monkeys to Africans has, however, rather oddly, evoked indignant cries of 'racism': sensitive persons have complained that Europeans were unjustly attributing the pandemic to African superstitions. But the suggestion of an African origin—correct or not—is no more racist than other statements about the geography of infections. Infection has often been brought from Europe to other continents. The indigenous inhabitants of Central and South America were at one time worse than decimated by appalling outbreaks of smallpox, measles, tuberculosis and other infections, carried by European invaders. (At the time, European physicians regularly bled their patients as a routine treatment—a harmful and superstitious act.) According to the Yanomamö of northern Brazil, a group threatened with extinction, white men cause illness: disease would not exist, they say, if there were no white men.

Another notorious sexually transmitted disease may have been carried to Europe from the West. Syphilis is believed to have been introduced into Europe by the Spanish sailors who reached America in 1492. From the sixteenth century it was a scourge in European and other countries. Like AIDS, it interferes with the functioning of the immune system; but it is a bacterial disease and can be cured with antibiotics.

Whatever the origin of HIV, the ways in which people are infected today are well known. Infection can result from either heterosexual or homosexual conduct when body fluids are transferred to the partner; from transfusion with infected blood; from sharing a needle which transfers infected blood when used for injecting a drug; and from transmission across the placenta of an infected mother to her unborn child. By 1991, about 10 million people were believed to have been infected with HIV.

As the infection spreads throughout the world, strenuous efforts are being made to educate people, especially the young, to forgo promiscuity or, at least, to use condoms. Both prevention and treatment are, however, handicapped by the geography of HIV. One prediction is of 40 million infected persons by the year 2000; the actual figure may be 110 million—2 per cent of the world's population. Many are likely to be in poor countries. At first, Africa had the highest figures of infection, but in the 1990s it began to spread rapidly in Asia. The nations most affected have health services ill equipped to deal with such a disaster: they cannot afford to import needed drugs or to educate people in methods of preventing infection. As a result, expectation of life will decline and millions of children will be orphaned.

HIV warns us of the always present possibility of new infections. Other viral diseases are emerging from obscurity. Eternal vigilance is needed to cope with these recurring but unpredictable hazards. The expression, hemorrhagic fever, is now, thanks to the new molecular biology, known to cover a number of formerly baffling, often fatal, conditions due to viruses. Most of the time, the viruses lurk harmlessly in various mammalian species, especially rodents. Recent outbreaks have resulted from ecological upheavals: cutting down forests and damming rivers can increase the contact of human beings with hardly studied species. To deal with these threats, we have to move from biological chemistry to ecology and back again.

The dangers have recently been sensationally shown by outbreaks of the Ebola virus of Central Africa. In 1995, it reappeared in Zaire (Congo) but the outbreak was contained. Where the virus hides between outbreaks is not known but infection is sometimes from monkeys. Most infected people die after a few days of bleeding and vomiting. No treatment exists. All that can be done is to isolate the infected population.

In 1989, a colony of crab-eating macaques (*Macaca irus*), in a military research institute in Reston, Virginia (near Washington DC), was found to carry a variety of the Ebola virus. It killed some of the macaques and monkeys of other species but, as it turned out, no people. General panic was prevented by keeping the event secret. The reader may try to imagine what would have happened if the virus had spread in the human population of Virginia.

Some threats, however, need no imagining, for ancient scourges are also returning, often in new forms. One is tuberculosis—the 'white plague' with which this chapter begins. The organism responsible is not a virus but a bacterium, *Mycobacterium tuberculosis*. Skeletons show that the disease has been present in human communities for thousands of years.

In its commonest form tuberculosis slowly destroys the lungs. It often kills people when young or in early middle age: famous examples are the playwright, Anton Chekhov (1860–1904), the composer, Frédéric Chopin (1810–1849), the novelist, Franz Kafka (1883–1924) and the poet, John Keats

(1795–1821). Hence it is a common theme in nineteenth century fiction and drama, such as Verdi's opera, *La Traviata*. Its worst impact has, however, been among poor people living in crowded squalor. During the 1930s, annual mortality in the United States was still about 90 000 and in Britain and France 50 000 or more. Yet, a few decades later, in these and other rich countries, tubercle had become a rarity.

That achievement was not the outcome of a single dramatic discovery. For many years, bacteriologists and cell physiologists had searched for a single agent which would kill the bacillus without harming the patient; and for a long time they failed: the bacillus has a resistant outer covering which protects it against 'magic bullets'.

Meanwhile, the enemy had been attacked by measures of public health. Much infection was prevented by requiring that dairy herds should be free of the bacillus and by pasteurisation of milk. Radiography of millions of people detected early signs of lung disease. A method of vaccination had been developed by two French physicians. Their names are acknowledged in the preparation called bacille Calmette-Guérin (BCG), which was used for large scale immunisation. Finally, three drugs taken together were found to be effective: one, streptomycin, is an antibiotic like penicillin; a second, PAS, is related to aspirin; and the third, isoniazid, is a derivative of nicotine. Prolonged and expensive treatment with all three is usually successful. In rich countries, by the 1970s, these measures, combined with improved living standards, made tuberculosis a rarity.

But elsewhere it has remained a menace. Early in the 1990s, in poor countries, it continued to kill three million people every year. And it is now reappearing in centres of wealth. In the United States, varieties of the bacillus resist the usual treatments: combined with HIV infection, they are killing increasing numbers of people. As a result of these and other changes, the US public health system has been described as falling apart.

Like the viral diseases just mentioned, HIV and tuberculosis therefore remind us of the unsleeping watch needed in the unending war against disease. Encouraging successes have come from combining cell biology with action by governments through national health services. But local measures are not enough. The world has justly been called a global village. An organism can move from any one city to any other in a day or two. International action is therefore needed as well.

It is not necessary to prophecy doom. We have the resources—moral, intellectual and material—to cope with our present condition. An English physician and writer, Richard Horton, has pointed to a parallel from history. Plague, the Black Death of the late middle ages, was a pandemic disaster but it also led to energetic improvements in care for ill people and in scientific medicine. The twenty-first century could see another such renewal.

THE MATERIAL OF HEREDITY: THE PROBLEM AND THE PEOPLE

It would indeed be quite wrong to end an account of cell biology with a digression on current dangers, for the most sensational and potentially valuable findings are still to come: they concern the cell nucleus.

In *Paradise Lost*, the poet, John Milton (1608–1674), calls sexual reproduction 'this fair defect of Nature'; and he asks why there is not 'some other way to generate Mankind'. Biologists ask a similar question. Reproduction among plants, fungi, animals and even protists is usually sexual. How has it come to be so widespread? The formation of a fertilised egg is not only more complicated than the asexual (or vegetative) method: it also requires two cells—which seems extravagant. An answer comes from our understanding of evolution. Each individual has genes from two parents. Advantageous genes may have arisen separately by mutation, and sexual reproduction can bring them together in one individual; hence new combinations of genes can appear. If the environment changes, new types, fitted to the new conditions, can quickly arise.

We do not know how sexual reproduction began. But, owing to a series of discoveries, which have justly been called the greatest achievement of science in the twentieth century, we know a great deal about how it is kept going. The seminal publication on the material of heredity was a paper, by Crick and Watson, in the journal, *Nature*, in 1953.

The problem arises from the facts of development. The fertilised egg of a many-celled organism grows into an adult

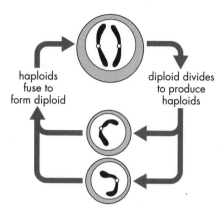

What sex means to a biologist. The nuclei of cells are of two kinds. The simpler ones (haploid) have a single set of chromosomes; each chromosome is different from all the others. In sexual reproduction two haploid nuclei from different cells fuse into one. This is fertilisation. The result is a nucleus with two sets of chromosomes—the diploid condition. In the figure, only one pair of chromosomes is shown. A fertilised cell (zygote) with a diploid nucleus has a set of chromosomes from each parent. Every chromosome is then matched by another, similar (homologous) one. Most of the familiar plants and animals, including ourselves, consist mainly of diploid cells produced, by many divisions, from the original fertilised egg cell.

Usually, the haploid nuclei which fuse to form a zygote nucleus are carried in special cells, the gametes: eggs (ova) produced by a female and sperms (spermatozoa) produced by a male. But it is possible to have sexual reproduction without sex: some protists come together and fuse nuclei; there is then no male or female, nor even sperm and egg.

closely similar to its parents. The contribution of the male parent is only the nucleus of a sperm. Both egg and sperm nucleus must therefore contain information ('instructions') which influence development. The material carrying the information must be copied every time a nucleus divides. Copying demands remarkable stability over generations. But occasionally, in any group of organisms, an exceptional feature appears and is reproduced in the offspring. The transmitted material must therefore be capable of occasional change, that is, of mutation.

Early in the twentieth century, the chromosomes had come to be seen as carrying many hereditary factors or genes, arranged 'like beads on a string'. But what were genes made of? Chromosomes consist mainly of water and two kinds of giant molecules, proteins and nucleic acids. Proteins are complex and diverse; hence, for a long time, the answer was assumed to be in protein chemistry. But the problem was solved only when attention was turned to the nucleic acids.

The story has a beginning in dilapidated laboratories in London. In 1928, a medical epidemiologist, Fred Griffith, made a disconcerting observation on pneumococci. These bacteria have genetically different forms. Griffith injected mice with dead cells of one type and living cells of another. Later, he grew cultures of the descendants of the living cells. Some were then found to be of the type which had been injected *when dead*. The dead forms had transmitted genetical information into live ones. Transformation of pneumoccoci affects only some of the live cells and is achieved only with difficulty. It was, however, confirmed by others. But Griffith himself never knew the importance of his finding. He died in 1941, when his laboratory was destroyed by one of the many bombs dropped on London by the Luftwaffe.

In 1944, in the Rockefeller Institute in New York, a further crucial observation was made by Oswald Avery and his colleagues. More than a decade of exacting experiments by this group pointed to sodium deoxyribonucleate as the transforming substance. Yet the nucleic acids seemed far too simple chemically to form genes.

The problem was finally solved by the exertions of many. Among them, first mention must go to Rosalind Franklin (1921–1958), an English member of a Jewish family with interests in publishing; although (or perhaps because) impressive as a person, she sometimes clashed with her male colleagues. Her death from cancer occurred before her contribution could be fully recognised. Her principal colleague, Maurice Wilkins, also a notable experimenter, extended his skills into the kitchen. Francis Crick describes how he inspected the pots on the stove when he was invited to dinner. Wilkins was also active in the Society for Social Responsibility in Science. Franklin and Wilkins, in London, were largely responsible for revealing the structure of the nucleic acids. They used a newly developed method, X-ray diffraction, which required much skill and very precise measurements.

Third is J.D. Watson, an American led to zoology by his interest in bird watching. In 1951 he went from Chicago University to Cambridge (England) and began his famous collaboration with Francis Crick. The usual adjectives, applied

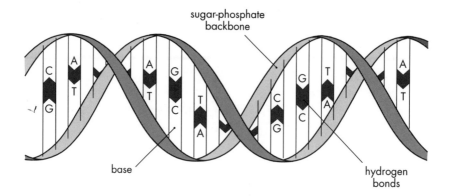

The 'secret of life': DNA. The DNA chain has a 'backbone' made of sugar units alternating with phosphate units. The sugar is deoxyribose. Attached to each sugar is another kind of molecular unit (a base) which contains nitrogen as well as carbon, hydrogen and oxygen. The bases project out from the backbone. DNA contains four bases: adenine (A), thymine (T), guanine (G) and cytosine (C).

That is one half of the DNA molecule. The other half is the same, but the bases are arranged in the opposite order. The halves are joined by the bases. Thymine always pairs with adenine, guanine with cytosine. A DNA molecule laid out flat would look rather like a ladder: the verticals would be represented by the sugar-phosphate backbones; the pairs of bases would form the rungs.

to him at that time, were 'brilliant' and 'brash'. His autobiographical book, *The Double Helix*, has been described by another distinguished scientist, P.B. Medawar, as a work of artless candour free from considerations of good taste.

Last, Francis Crick, intellectually the most formidable of this group, comes from an English middle-class family associated with manufacture. Like his colleagues, he was far from matching the conventional myth of the cold, detached scientist. He is said to have gone into a pub one day and 'told everyone we'd discovered the secret of life'. Years later, his wife told him she had not believed a word of it: 'you were always coming home and saying things like that, so naturally I thought nothing of it'.

A short account of such distinct personalities, and the exciting character of their discoveries, can give a misleading impression of research. First, it overlooks the contributions of others. Modern biological findings are the product of an international community which includes many hardly known scientists; still less well known is the essential corps of research assistants, graduate students and technicians. Crick's autobiography, *That*

Mad Pursuit, goes part way to remedying this neglect. Second, research is not a uniformly coherent, 'logical' process: blunders, false starts, blind alleys and sheer stupidity all play a large part. Third, research is rarely thrilling from day to day, even for those who wish to do nothing else. Crick writes of 'the gradual accumulation of experimental facts'; and, of one investigation, 'It is difficult to convey just how tedious and complicated such an experiment is'.

THE DOUBLE HELIX AND THE CODE

The first major group of discoveries concerned the structure of DNA, opposite above. The most important feature of DNA is the sequence of pairs of bases which form the rungs of the helical 'ladder'. The order in which the four bases (A, T, G and C) are arranged varies. This discovery led to a crucial hypothesis: that the order makes the 'genetic code' which determines what proteins are made in each cell.

Proteins are chains of amino acids. (They also have side chains.) Twenty amino acids make up virtually all proteins. If the base pairs in DNA are the genes, they must determine the order of the amino acids in the protein chains. The order determines the properties of each protein. Chromosomes were already known to contain strings of genes which sometimes mutate. According to the hypothesis, mutation of a gene should consist of a change in the arrangement of base pairs.

With that hypothesis, the main features of the code were deciphered with surprising speed. A 'triplet', consisting of three bases (a codon), codes for a particular amino acid. A gene consists of tens of thousands of identical codons. The sequence of codons in the DNA molecule is copied in the RNA, and the RNA specifies what proteins are synthesised. Hence follows what is sometimes called the 'central dogma' of modern genetics:

$$DNA \rightarrow RNA \rightarrow protein.$$

This statement is, however, misleading in two ways. First, the DNA is not independent of the rest of the cell: it interacts with the proteins; it does not copy itself but is copied as a result of this interaction. Second, much of the DNA in each cell is 'noncoding': it seems to have a role quite different from that of the code. (More about this in chapter 10.) RNA does not include copies of the noncoding regions of DNA: it matches exactly the proteins synthesised in the cell.

Engineering the Genes: Centaurs Next?

Much of the early work on the nucleic acids was carried out with minimal facilities in cramped quarters. The workers hoped above all to contribute to the understanding of nature. This disinterested research has now, however, led to massive practical applications, for gene engineering has become a modern version of an ancient practice. For thousands of years, farmers and stockbreeders have improved crops and domestic animals by selective breeding. Some familiar foods are from selected mutant plants: that delicious fruit, the nectarine, is a mutant peach. As well, varieties have been crossed and have assembled, in the resulting stock, the useful features of both. The mule, derived from mating horse with ass, combines characteristics which make it an admirable draft animal. Nearly all corn (maize, *Zea mais*) now grown in the United States is hybrid and is much more productive than the original varieties.

Hybridisation by traditional methods is possible only within species or between closely related species. It also takes a long time and often fails. Today, experimenters take advantage of the universality of DNA: they can rapidly transfer genes between forms as separate as bacteria and flowering plants or mammals. This is called transfection. Any species can be a source of useful genes.[1]

Botanists are now achieving transfection on a large scale. Some wild forms differ from closely related crop plants by possessing immunity to disease or to attack by insects. If the difference from domestic forms is genetically determined, the trait can sometimes be transferred, via a microorganism, to another plant. Successful transformation of food plants is most likely when the new trait is a clearly defined chemical property, such as secretion of a fungicidal or insecticidal substance, coded for directly by a single gene.

An interesting but probably distant prospect is of engineering cereal crops, such as rice or wheat, so that, like legumes, they can fix nitrogen. Cereals produce much more food per hectare than do legumes but they deplete the soil. The bacteria on the roots of legumes, such as beans and soya, enable these crops to leave behind, in the soil, plenty of nitrogen compounds. The ability to fix nitrogen is not, however, a simple feature switched on by a single gene. Moreover, the outcome of success

Manipulating DNA

Until recently, the DNA molecule seemed unmanageable. From a chemist's point of view, one DNA is like any other: each consists of exactly the same units—the four bases together with a sugar and phosphate. The molecules are enormous: each contains millions of base pairs. Yet today DNA is easily analysed. We owe this to enzymes first described, in 1962, by Werner Arber of the University of Geneva. The restriction endonucleases, which are present in many bacteria, cut DNA molecules into short lengths. Evidently, they defend the bacteria from destructive viruses (bacteriophages): the enzymes chop up the virus DNA.

More than a hundred restriction enzymes, extracted from bacteria, are available in a pure form. DNA taken from any organism can be divided into fragments of standard length by one of these enzymes. The different kinds of fragments are separated by an elaborate technique (electrophoresis). Base sequences which code for particular proteins can then be isolated and joined to the DNA of microorganisms (bacteria or yeasts). The product is recombinant DNA; and, when the microbes multiply, it is copied (cloned) in the same way as normal DNA. The copied DNA can be injected into the nuclei of fertilised animal eggs or into plant cells in tissue culture. The end result is a transgenic organism—a hybrid with genes derived from another species.

might itself present problems: fixing nitrogen uses much energy; hence the productivity of a nitrogen-fixing cereal could be disappointing.

Not all transfection is concerned with food crops. Growing decorative flowers is a giant industry which benefits from the creation of new forms. One novelty, long desired, is a blue rose. The expression, blue genes, sounds like a disreputable pun but, in 1991, it came to have a commercial significance: an Australian enterprise has patented a method of transfecting roses with foreign DNA from petunias and hopes to put blue roses on the market.

The imagined scope of these methods is summed up in a witticism by a leading molecular biologist, Sydney Brenner: 'Genetic engineering', he suggests, 'is being able to build a centaur'. The creation of such a mythical being—half man, half horse and always male—would perhaps not be universally welcomed. But what about the *fur* tree on page 28?

Anxiety about such improbable forms has appeared in fiction. In an entertaining crime novel, *Green Grow the Dollars*, by Emma Lathen, most of the characters are involved in agribusiness. One exclaims:

> . . . the media boys have distracted us with all their hype. They've scared us to death with talk about creating new forms of animal life, cloning human beings . . . And while we've been bird-dogging the zoologists, it's the botanists who've been having a field day . . . We're probably all going to be killed by a man-eating petunia.

As it happens, a gene from petunia plants has been transferred to bacteria, and from them to tobacco plants whose resistance to disease has been thereby improved. But, despite the many existing carnivorous plants, man-eating petunias seem unlikely.

Other problems arise from the use of genetically engineered foods. Will they contain natural poisons or substances responsible for allergies? Some may incorporate pesticides which could injure people as well as pests. Genetically engineered organisms can also be ecologically dangerous. Human invaders of new lands have often brought strange species with them, from microorganisms to large mammals. When the introduced species have flourished, the outcome has often been a disaster. A genetically modified form, released accidentally or deliberately anywhere in the world, could become a destructive weed, a pest or a pathogen. Governments therefore need to screen new organisms as they do new drugs.

What of the many enthusiastic predictions about the medical impact? Gene therapy is a new kind of treatment by 'magic bullets'. Some rare illnesses and defects are directly related to single abnormalities of the DNA, that is, to single mutant genes. One (already mentioned) is severe combined immunodeficiency. To treat such conditions, use may be made of the ability of viruses to penetrate cells. A population of viruses is first infected with the required gene. White blood cells are then painlessly removed from the patient, exposed to the virus and returned carrying the gene. Amazing to relate, the method works:

children with SCID, otherwise condemned to a short and drastically restricted existence can, after such treatment, become able to live a normal life.

Cystic fibrosis too is, or seems to be, genetically simple. In this case, the main trouble is in the lungs. Treatment may therefore be attempted by applying carriers of the replacement gene directly (*in situ* therapy). The same method can, it is hoped, also be used against cancers: preparations of viruses, injected into malignant growths, will carry genes which kill the malignant cells. Such methods are not easy to apply: the viruses may themselves be injurious; and the genes, once they reach the patient's cells, may not promote the synthesis of enough of the effective protein.

Nonetheless, the new knowledge is already highly profitable. One outcome of modern genetics is the market in human and other genes. An American was successfully treated for a form of leukemia. The treatment included removal of his spleen. Cells from the spleen were grown in tissue culture and were patented. They therefore became private property and were sold for US$1.7 million by a research institute. Drug companies are now involved in a struggle to prevent the use of human DNA sequences except for their own profit. They and other organisations are also trying to establish exclusive rights in sources of genetical material which range from trees in tropical forests to the hundreds of small tribal groups whose survival is precarious. Because these activities are dominated by market forces, the enthusiastic publicity they receive often resembles the work of advertising agents.

The preceding paragraphs may seem unnecessarily dampening. They do, however, match the present state of the subject. In 1996, in their *Human Molecular Genetics*, two American geneticists, Tom Strachan and Andrew Read, write:

> . . . molecular genetics is poised, at last, to deliver novel treatments for human disorders. Exciting though this prospect is, the limitations of the current technologies are apparent . . . Even now, gene therapy has not *cured* any patient. Instead, current gene therapy trials are providing forms of *treatment* for some disorders: . . . but the effects are temporary, and treatments have to be repeated at regular intervals.

The medical scope of engineering genes has therefore still to be mapped. The current stories in the media often arouse unjustified hopes and distort real and impressive advances.

GENE ACTION IS NOT SIMPLE

Concentration on DNA also obscures more fundamental problems. Much of the success of genetics has depended on studying genetically simple features, such as blood groups, which are, or seem to be, independent of environmental variation. It has therefore become easy to believe that many important traits are switched on or off by single genes. This is quite wrong. A large gap usually exists between the direct action of a gene and a trait in which we are interested.

Among the many possible examples is an uncommon condition of infancy, pyloric stenosis. About three in every thousand infants, at the age of five days or later, are found to have the opening of the stomach into the intestine narrowed by excessive growth of a muscle (the pyloric sphincter). This interferes with feeding but it can be successfully treated by diet or surgery.

Genetic variation exists in the liability to develop pyloric stenosis. Since the human species is genetically very variable, that is to be expected. It does not, however, justify the statement, implied in some medical writings, that the condition is genetically determined. It is not known how the condition develops. Since, however, it is four times commoner among boys than among girls, one factor is evidently the sex of the fetus: conditions in a male fetus differ from those in a female and the difference influences the growth of the muscle. More important, pediatricians find the condition to be influenced by the feeding regime offered as soon as the child is born. A long sequence of events evidently intervenes between the direct effects of the relevant genes (which are part of the genotype) and the overgrown sphincter (the phenotype). At each stage, environmental variation may influence the outcome.

The belief in simple connections between genes and characters is especially misleading when applied to our socially most important features. These are not single, distinct traits. Health, intelligence and moral conduct are influenced by differences in many genes, of which the actions are usually unknown. Much more important, the effects of the genes vary with conditions of upbringing. At every stage in development, from moment to moment, the growing organism is interacting with a varying environment; and the form of each interaction depends on the outcome of earlier interactions. This process is indescribably complicated—which is why it is never described and rarely even acknowledged. The extreme of intricacy is reached in human development, for the conditions which we and our children experience are often the products of deliberate, sometimes intelligent, choice. So, in the next chapter, we come to a central theme of human biology: how do heredity and environment interact?

Chapter 4

The Authentic Gene

Allow not nature more than nature needs.

SHAKESPEARE

When a character in a novel by P.G. Wodehouse was asked whether he believed in heredity, he replied, 'Of course; that's how I got my money'. The primary meaning of words such as inherited is that implied by Wodehouse's young idler: the words belong to law or custom. If you inherit a desk or a diamond from a relative, the object is handed over unchanged; but that does not apply if you 'inherit' your mother's nose or your father's temper.

Another source of confusion is the use of blood for ancestry. Even today, the reader can hear the expression 'bonded by blood' during a discussion of human relationships. A person whose parents are of different types may be said to be of mixed blood. Such expressions encourage the notion of biological inheritance as a blending process like combining paints. This is supported by (for instance) the result of the mating of a black person with a white: the children are usually brown.

In biology heredity refers to processes much more complex than acquiring a legacy or mixing pigments. In sexual reproduction parents transmit to their offspring large numbers of genes assembled in small numbers of chromosomes. A feature such as colour or nose shape or fiery temper is an outcome of a long process of development from egg to adult. Genetics, the science of variation and heredity, is about the relationships between the genes in the egg's nucleus, on the one hand, and the differences between individuals, on the other.

Since modern genetics began, pronouncements supposedly based on genetical science have influenced public attitudes and the policies of governments on education, health, the roles of the sexes in society and on race relationships. Here are statements which have been made (some still are): pauperism and criminality are inherited traits; intelligence is determined by a person's genes; insanity and homosexuality are genetically determined; black people are genetically incapable of contributing to civilisation; people with almond eyes (the epicanthic fold) are more intelligent than Europeans; women are fit only for home making, not for public affairs or the professions. All these assertions are false or misleading. Yet, since 1937, large funds for propagating such ideas have been available from the Pioneer Foundation, an American organisation dedicated to 'race betterment'. Millions of dollars have endowed ostensibly scientific studies of which the methods and conclusions are seriously flawed.

Several odd but popular beliefs appear in other chapters: that character is shown by bumps on our heads; that intelligence can be explained by conditional reflexes; that the conduct of insects and other animals can explain human social life. Past medical prescriptions have included necklaces to ward off evil spirits and bleeding patients to treat almost any ill. Today, modern genetics provides an excuse for similar magical practices: fragments of DNA, cloned from sports champions or pop stars, are worn in amulets. They have been likened to the splinters, alleged to be from the true cross, cherished in the European middle ages.

MENDELISM: PINK FLOWERS AND BLUE BLOOD

Truly scientific genetics began with the study of genetically simple situations, especially those offered by flowering plants, insects and mice. The story begins in 1866 with the Austrian priest, Gregor Mendel (1822–1884), a member of a peasant family who studied natural science, especially physics, in the university in Vienna. He taught science in Brünn but failed the teaching certificate. His most important work was on the many, easily crossed varieties of garden peas (*Lathyrus*). His central conclusion was that, in sexually reproducing organisms, hereditary factors (now called genes) are present in pairs; they influence all development and are transmitted from parent to young as separate units: no blending occurs.

Mendel's famous paper was a version of lectures delivered to the Natural History Society of Brünn. His method was novel. Previously, hybridisers had described whole plants. Mendel concentrated on pairs of distinct, contrasted features: the length of the stems (long or short); the shape of the seeds (smooth or wrinkled); seed colour (yellow or green); and others. He also recorded the *numbers* of plants with the different traits. He was a pioneer of quantitative analysis in biology. During Mendel's life his work was not recognised; but in 1900 two German workers and a Netherlander separately reported similar findings on other species. Yet, even then, 'Mendelism' was not generally accepted. Some leading geneticists were occupied not with unit features but with measurable traits such as human stature or yields of beef or grain.

Later it was shown how continuous variation too can result from the operation of separate genes. The mathematician and evolutionist, R.A. Fisher, who achieved this, once suggested that Mendel's findings were questionable, because they were too good to be true. They were imperfectly presented and the original notebooks have not survived. Soon, however, several plant breeders separately repeated Mendel's experiments and reported similar results. Yet journalists still sometimes revive the tale that Mendel's figures are bogus.

MENDELIAN RATIOS

The flowers of the snapdragon (*Antirrhinum*) may be red or white. Cross them, and all the offspring are pink, which looks like blending. But now breed from the pink snapdragons: about half the next generation are again pink; but in this second generation the two original colours reappear undiluted: a quarter are red and the other quarter, white. The differences depend on a single gene which exists in alternative forms (alleles). The pink snapdragons owe their colour to having one gene from a red parent and one, slightly different, from a white.

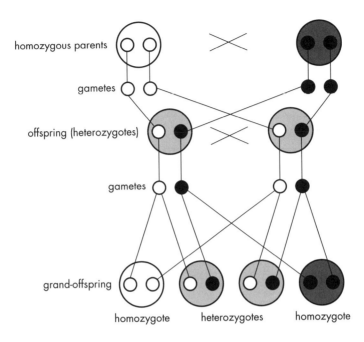

'Mendelian' inheritance in its simplest form. The diagram corresponds to the case of the flower colours of *Antirrhinum*, but it represents a general and fundamental discovery about biological inheritance. It has also led to ghastly misunderstandings because, in fact, genetics is not simple.

Relationships such as those of snapdragon colour help us to understand heredity but they are quite unusual. One complication is dominance. Vinegar flies (*Drosophila melanogaster*) have bristles on the head and body; in the laboratory, a few turn up with short bristles ('stubbloid'). If a normal fly is crossed with a stubbloid, the offspring all have normal bristles; but stubbloid reappears in about one quarter of the *next* generation. In this case the presence of one normal gene prevents the alternative gene (the allele) from influencing bristle development. Only the 25 per cent of flies—those with two copies of the stubbloid gene—show its effects. If two stubbloid individuals are crossed they breed true: that is, all their offspring are stubbloid. Normal bristle length is dominant and stubbloid, recessive.

One result of dominance is that individuals with the same appearance may have different genes. This is an example of the crucial distinction between genotype and phenotype. One cannot tell, from seeing the phenotype—that is, the actual organism—whether a fly with normal bristles has one stubbloid gene or none.

Modern genetics was fully launched only when the details of sexual reproduction had been discovered (see illustration below). A sperm and an egg each has one set of chromosomes, hence one set of genes. The fertilised egg has two sets, one from each parent. Most cells of many-celled organisms have such a double set. By this means each parent makes a similar genetical contribution to each offspring.

The correspondence of the chromosome system with the results of breeding is seen in human sex-linked traits. A woman has two X chromosomes in each ordinary cell. A man has only one, together with a smaller Y chromosome. Some genes on the X chromosome of a man are therefore not paired. One result is seen in red blindness: the inability to distinguish red from green occurs in more than one per cent of men but is rare among women. The sexes differ in this way because the relevant gene is on the X chromosome. A man can be red blind with only one abnormal gene (acquired from his mother); but for a woman to be red blind she must have two of the abnormal genes, one from each parent, which happens much less often.

Another example is hemophilia, in which the blood clots very slowly. Hemophiliacs may bleed severely from minor, especially internal, injuries; they usually die young. Nearly all hemophiliacs are men. Hemophilia is transmitted by a woman, herself normal, who carries a gene which is likely to kill half her sons. Each of her daughters has an even chance of carrying the gene. Queen Victoria of England was a carrier: one of her sons, at least three grandsons and six greatgrandsons were hemophiliacs, including members of the ruling families of Spain and Tsarist Russia.

Victoria's children were born during the period when modern eugenic ideas were beginning to be thought of. If the eugenic movement had already got under way, and the proposals of extreme eugenists had been accepted, it is possible to imagine, after the birth of her first hemophilic son, a visit to Buckingham Palace by a State Inspector of Eugenics, and a demand that Her Majesty be sterilised. Or would persons with blue blood have been exempted? (More on eugenics later.)

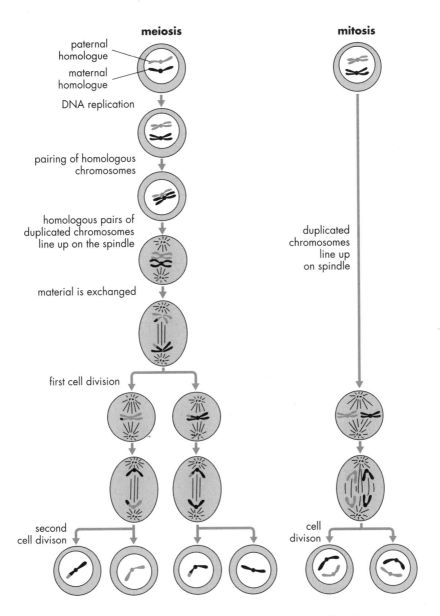

The two kinds of nuclear division. Only one pair of similar (homologous) chromosomes is shown in each sequence. In mitosis (on right) the complete set of chromosomes in a nucleus is duplicated; two daughter nuclei are formed, each with the same number of chromosomes as the original nucleus. Usually, mitosis is followed by division of the cell. (Compare page 9.)

Meiosis, in the formation of sperms and ova, entails only one division of the chromosomes but two divisions of the nucleus. The original germ cell also divides twice. The result is four nuclei, each with the half (haploid) number of chromosomes. By this means, in a testis, each germ cell gives rise to four haploid sperms. In an ovary, only one of the four cells becomes an ovum: the other three hardly develop.

Genetics is Not All Mendel: Many Genes, Much Variation

A reader who stopped now would have a seriously distorted view of genetics. I have said nothing of mutation, of the multiple action of genes or, above all, of the influence of the environment on variation.

Mutations are of several kinds. We are now concerned with single gene, or 'point' mutations. As an example, hemophilia can appear suddenly in a family with no previous record of abnormality. The continued existence of hemophilia has, in the past, been due to a steady incidence of mutation.

The effects of mutation can be seen in natural populations. Many animal species occasionally throw up a white (albino) individual. When the parents are normal, this is usually a result of both carrying one copy of a particular mutant gene. From such a mating, one quarter of the offspring, on average, lack normal pigmentation. (Hence albinism is said to be recessive.)

Albinism in wild populations illustrates a general rule: in natural conditions, to be abnormal is usually disadvantageous. Most small mammals are unobtrusively coloured and are seen with difficulty by hawks or foxes. White ones would be easy marks and unlikely to live long. Human albinism, too, is usually recessive but it is disadvantageous for different reasons: the skin, hair and eyes contain no pigment; the skin has therefore no protection from sunlight and bright light interferes with vision.

Another, fundamental rule is that, when we speak of advantage and disadvantage, we must say what environment we are talking about. The albinism of rats and mice is favoured in a laboratory or a pet shop. Even human albinism has been advantageous in special circumstances. Among the Hopi American Indians, albinos are numerous. While the healthy men were in the fields the albino men stayed in the villages with the women; they therefore sired more children, many of whom were themselves albinos. In these conditions, the albinos were biologically fitter than the others. This is a rather odd example of natural selection in action.

Unfortunately, phenomena such as albinism make it easy for nonspecialists to get the impression that genetics is all about the effects of single pairs of genes and that each characteristic can be switched, from one state to another, by a change in a single gene. In fact, few important features are influenced in this way: often they vary smoothly and have to be measured on a continuous scale. In ourselves, think of stature, skin colour and intelligence quotient; or, in domestic species, yields of grain, fruit, beef and wool.

Much of the variation is due to environmental factors, such as the food available and the climate. To identify genetical influences, one method is to breed a mixed stock selectively. Cats, cattle, camels, pigs, dogs, poultry and others were long ago drastically changed from their usual forms without benefit of Mendel. Domesticating a species always entails selection: when plants or animals come in from the wild, only some breed. Domestic animals are selected for docility and often for the ability to tolerate crowding. Later, they can be further altered. So we have giant draft horses for pulling carts or plows but quite different breeds of horses capable of running several kilometres at 45 kmh while carrying a jockey. We also have trees with fruit larger and sweeter than anything found in the forest. All such changes are influenced by at least several genes.

When both genotypes and environments vary, how do we distinguish their effects? Sometimes environmental action is obvious. If I described the oak, *Quercus robur*, as a low, straggly shrub, I might well be thought astonishingly ignorant for, if acorns are planted in well watered ground near sea level in (say) the south of England or France, some eventually become magnificent trees. But, if similar acorns are planted at 400 metres on rocky hillsides further north, those that survive produce only low shrubs. An oak tree can be tall *or* a mere shrub, according to the environment. Similarly, a herd of Jersey cows would give large amounts of creamy milk in the first environment but little milk in the hills.

Important variation, however, usually reflects the multiple action of both genes and environments. Over the page are shown plants of a single species. Those in the top row are of a variety that usually grows at high altitude; the middle type grows in meadows lower down; the bottom row shows a coastal form. The three varieties probably differed in many genes. Cuttings of each type were planted at each of three altitudes. The three environments therefore differed in temperature, rainfall, exposure and other features. Each plant illustrates the combined effects of genetical and environmental differences.

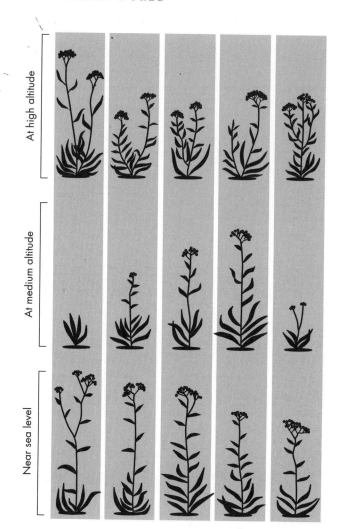

Gentle reader, note well! The interaction of genes and environments, illustrated by five genetically different varieties of a plant. How well each plant grows depends both on its genes and also on the conditions in which it is cultivated.

Study of this picture shows why the interaction of environment with heredity causes difficulty. 'Heredity' in ordinary speech refers to *likenesses*: individuals resemble their parents or grandparents and so on. But genetics is about *differences*, such as those between the plants. It is therefore better not to speak of *characters* as genetical (or inherited) *or* as environmentally caused: they are always both. It is very difficult to avoid this customary way of speaking; yet the principle is of the utmost importance, especially—as we now see—in its application to people.

Questions on Human Breeding

The hope of breeding a superior race of human beings has entranced thinkers for at least two and a half millennia. The modern eugenic movement belongs, however, almost entirely to the twentieth century. Inspired by the new genetics, it was said to be scientific; yet it arose from prescientific presumptions about heredity. In nineteenth century fiction a common theme was the emergence of excellent character in a person who has suffered an appalling upbringing. In *Oliver Twist*, by Charles Dickens (1812–1870), the hero spends his first years in the squalor of a workhouse with boys of low degree. His true parents, however, had been superior persons. Hence, from the first, he speaks excellent English and he later develops as a splendid fellow. Evidently, his parents had provided him with genes for being articulate and good, regardless of early experience.

The eugenists talked enthusiastically about genes for almost every important human trait, from administrative ability to zest for life. Slaves were said to have a gene for running away; sailors, for running away to sea; poor persons, for criminality; professors, for intelligence; unwanted immigrants to the United States, for imbecility.

The genes were invented. Their importance was in the social programs they served: they supported 'racial purity' and they implied that attempts to improve the condition of the poor were futile. Poverty was held to be an inherited condition. As one example, pellagra is an extremely unpleasant disease, formerly widespread among poor farmers in the United States, which affects the brain, the skin and the bowels. The eugenists said that pellagra was inherited and, during the 1920s, American governments used this story to support inaction. It had, however, in 1916, already been found to be due to shortage of a vitamin, niacin, in the diet. Pellagra began to be dealt with effectively, by improving diets, only in 1933.

The early eugenists emphasised breeding for excellence (positive eugenics). In the 1930s such programs were advocated especially in Nazi Germany. Today, proposals for negative eugenics are more prominent. We now have reliable information about the genetics of many narrowly defined features, some of which are defects. Can they be eliminated if certain people have no children?

Above I give examples of sex-linked traits. They include red blindness—a minor inconvenience which nobody wants to eliminate—and hemophilia—a severe handicap. The latter offers little scope for eugenic action. Few hemophilic men have children. A woman who carries the hemophilia gene is not, as a rule, identified until she has a hemophilic son; she could then decide to have no more children. New methods have recently isolated mutant forms of the relevant gene; perhaps therefore some carriers will soon be identified in early life. But, even if all such women decided to remain childless, the gene would still, as before, be kept in the population by mutation.

For effective action, dominant conditions may seem more promising. They are transmitted only by affected persons. Each child, born to such a parent, has an even chance of developing the condition. An example is brachydactyly, or short fingers, due to imperfect growth of the middle bone in each finger. This does not prevent a normal life. Other traits are more serious: in split hand (ectrodactyly) each hand has only a thumb and one distorted finger. This does not entail bad health. If all such persons refrained from having children, ectrodactyly would decline; but it would still appear occasionally, again owing to mutation. Should ectrodactylous people have children? A celebrity in Los Angeles, herself ectrodactylous, has recently taken the risk and her child was ectrodactylous too. She has been fiercely (I think unjustly) criticised for her decision.

Worse than ectrodactyly is a form of dwarfism (chondrodystrophy): the skeleton does not grow normally and the limbs are short and may be deformed. About one such person is born in every 10 000 births. Only a few marry and have children; hence little scope exists for eugenic action: the condition reappears regularly owing to an unusually high mutation rate.

To discover these 'Mendelian' conditions took long, painstaking examination of people and pedigrees, followed by statistical analysis. Many difficulties arise. The effects of single mutant genes often vary. Lamellar cataract, a dominant defect of the cornea of the eye, can cause blindness but can be successfully treated in early life. Often, however, it consists of an opacity so small that it can hardly be detected and needs no treatment.

Recessive defects too at first sight seem genetically simple. They nearly always appear when the parents, both normal, each carries one copy of a mutant gene. Each child then has one chance in four of developing the condition. Half the children, on average, carry one copy of the gene but are themselves normal. Rare recessive conditions include a fatal skin disease, xeroderma pigmentosum; a severe glandular condition, cystic fibrosis, already mentioned, which especially affects the lungs; and a form of mental defect, Tay-Sachs disease, which kills early in life.

Descriptions of such genetically simple traits may reinforce the idea of genes as destiny. Even scientists often write and talk of 'genes for' certain ills. In doing so, they commit two serious errors, both illustrated by cystic fibrosis. First, such expressions ignore the environment: cystic fibrosis can be treated with drugs, antibiotics and physiotherapy. Second, there is no one 'cystic fibrosis gene': in the relevant DNA sequence 360 distinct mutants have been identified. The effects of the mutants range from mild to very severe.

A clear case of an important environmental effect comes from a much studied condition, phenylketonuria (PKU). Although it can be detected at birth, its ill effects appear gradually: the child grows up mentally retarded. The defect is due to the action of an amino acid, phenylalanine, on the developing brain. Clearly, PKU is an 'environmental' condition, due to a poison, phenylalanine.

Yet PKU appears in the textbooks as a genetical or inherited disease. It is in fact another recessive condition: children with PKU have parents both of whom carry one copy of a rare gene. Phenylalanine is a component of all normal diets and is a poison only to such children. Until recently, a phenylketonuric infant was therefore condemned to mental deficiency. Today, in rich countries, the newly born are routinely tested for PKU and those with the condition are given a special diet during their first years. They may then develop normally.

So here is a case of an 'inherited' disease which can be effectively treated. Diseases, therefore, like other traits, should never be described as inherited: doing so confuses a biological phenomenon with the legal process by which an object or a name is passed from parent to child; it disregards the fact that development is influenced not only by genes but also by a varying environment.

Environmental influences are of many kinds. In another notorious recessive malady, sickle-cell anemia, the blood hemoglobin is abnormal; as well, brain damage, malfunction of the heart and abnormalities of the bones may occur. These are multiple effects on development due to a single gene (pleiotropy). The condition causes about 100 000 deaths annually, most of them before the sufferer is old enough to have children. Hence the numbers of the abnormal gene ought to be kept down by the deaths of those who have two copies (the homozygotes). Yet, in some parts of Africa, the gene is common. In these regions a severe form of malaria is prevalent. People with *one* copy of the sickle-cell gene (heterozygotes) are resistant to malaria and are otherwise healthy: they are adapted to the local conditions. Survival of a population in a malarial region is therefore aided by a high incidence of the gene.

When such a group enters a different environment, the incidence of the abnormal gene changes. Africans descended from slaves have, after several generations in the United States, a lower incidence of sickle-cell anemia. There they are not exposed to malaria: the gene is no longer advantageous and is being slowly eliminated by the early deaths of those with the disease. In this case a 'eugenic' effect is due to changing the environment of a whole population. When malaria is finally overcome, the disease will decline in Africa too.

Despite all the facts, we are told almost daily of newly discovered 'hereditary' conditions. So, to show what can result from fixation on genes, here is one more case history. Recently, two American geneticists announced the 'discovery' of a gene for attending medical school. They had asked a large number of medical students which of their relatives had studied medicine. The findings were then analysed by a computer programmed to detect Mendelian ratios. The result: attending medical school is a recessive condition. The authors of this research are, however, serious geneticists. They published their finding to show how narrowly based assumptions can lead to absurdity. In this case the major omission was the family environment. Choice of medicine as a career is much influenced by close relatives who are already physicians. The performance of the computer also illustrated a well established law of computer science, abbreviated as GIGO: garbage in, garbage out.

THE HUMAN GENOME: NATURE AND NURTURE

In mid century, with the rise of genuine human genetics, the eugenics movement lost impetus. Genetics can be usefully applied to human traits only when the genetical effects are simple—and often, as we have seen, not easily even then. The causes of variation in most of our important features (good and bad) are far from simple. Yet eugenic ideas have now been revived. In the previous chapter I describe the prospects of gene therapy. These are being encouraged by a massive research program under an international Human Genome Organization or HUGO. The human genome is the complete set of genes on our 46 chromosomes. The number of genes is estimated at about 70 000. Technical advances make it possible to identify each one. That is HUGO's objective.

One prospect is of early diagnosis of more ills and hence more frequent prevention. An example is osteoporosis, a bone disease common among elderly women. (It also afflicts men.) Fracture of the thigh bone is a frequent outcome. A good diet, taking plenty of exercise and refraining from smoking greatly reduce its likelihood. Hence it is largely preventible. (And, if osteoporosis develops, it can be treated by diet and exercise and by hormones.) Some people, perhaps 16 per cent of Europeans, with two copies of a recently identified gene, are believed to be especially liable to osteoporosis. In rich countries with good health services, a simple test for the gene may become available. People at risk may then be identified early and warned.

Such findings, concerning other infirmities, have been made; more will be. The attitudes of the people concerned will then often be crucial. An infant born with Tay-Sachs disease can live only a few miserable years. The parents too suffer severely. In some populations the condition is common. Diagnosis before birth allows abortion. Hence, among people most at risk, the incidence of the condition has greatly declined and much misery has been prevented. Such action is resisted only by the most callous and doctrinaire opponents of abortion in any circumstances.

Tay-Sachs disease is a clear case. Other ills are not. Huntington's disease is a dominant condition in which changes in the nervous system cause progressively worse involuntary movements and other signs. Onset is often at a late age, hence some people with the gene must die before the signs develop. When the signs do appear, the sufferer may have led a good life and he or she may have had children. Each child has an even chance of inheriting the gene. Today, people with the gene can be identified early. They can then be warned about what may happen (but not when); and they can decide whether they will risk having children with the same gene.

Problems, however, arise even in such an apparently simple case. One is that different mutant genes can have the same effect. A person may be suspected of having the relevant gene but be told that he or she does not carry it. Yet, after that, the disease may develop. Such false negatives occur, with present knowledge, in about one per cent of cases. A similar problem has now arisen with Alzheimer's disease which also comes on late in life and has been said to be related to a mutant gene; about one prediction in every ten is wrong.

Additional dangers arise from trade. A certain mutant gene is believed to confer a high likelihood of developing breast cancer. This prediction, however, is valid only if the woman's family has a history of such cancers. Perhaps the cancer requires the presence of other genes; if so, it is another example of the complexities of gene action. The extra hazard comes from salesmanship: tests for the gene are being energetically sold, like automobiles or detergents. Hence many operations (mastectomies) have been performed without need.

Worse, in the United States, which has no comprehensive national health service, healthy people carrying mutant genes have been prevented from taking out health insurance. Some have been refused jobs, on the ground that they are genetically diseased. Alarm at this atrocity is fully justified, for we all carry some unwelcome genes.

The human genome project has therefore aroused much anxiety, even hostility. Yet, for some people, mapping the human genome has had a dazzling effect, like that of early genetics. In 1989, a prominent biologist and leader of HUGO stated: 'Now we know, in large part, that our fate is in our genes.' The attentive reader will quickly identify a major error.

Development is not fixed by the genes: it happens in a series of environments. Each trait is part of the phenotype: it is an outcome of growth in variable conditions. As PKU and sickle-cell anemia show, even when a feature is genetically simple, interactions with the environment can be quite intricate. People who say that genes are overwhelmingly important, or that the problems of heredity and environment can be disregarded, therefore make real difficulties more difficult.

Nor is it enough to say that heredity and environment have equal weight. Every trait in which we are interested must be examined separately. Concerning each one, we should ask: is there evidence of significant genetical variation and what, if any, are the relevant environmental influences?

Despite modern genetics and HUGO, answers to the second question are usually more important. About three per cent of human diseases are said to be related to single genes (in the sense that cystic fibrosis is so related). They are all rare when compared to the worst scourges we see around us. Some of these ills can be remedied by treatment but, in the long term, the most important findings concern prevention. Above, I mention Alzheimer's disease. From reports in the press, many people must have the impression that a 'gene for' the malady has been found. Now, however, environmental effects too are being revealed. One finding is of an adverse effect of heavy smoking; another concerns atmospheric pollution.

Other prominent infirmities are midlife diabetes and diseases of the heart and blood vessels. They are common especially among people whose diet is unwise, who take little exercise and who smoke. Genetical variation in susceptibility undoubtedly exists, but it is unimportant compared with lifestyle.

Some widespread ills remain baffling. A question which can cause much anguish is the inheritance of insanity. Among the common mental disorders are those called schizophrenia. General features include withdrawal from reality, inertia and a belief in threats or persecution by imaginary enemies. (Schizophrenia should not be confused with the entirely different, rare phenomenon of dual personality or dissociation.)

The condition is mildly familial: a person who has a near relative diagnosed as schizophrenic is slightly more likely, than are other people, to suffer a similar disorder. Yet, even when both parents are schizophrenic, only about 40 per cent of their children are similarly diagnosed. And, still more surprising, the most careful studies have disclosed a concordance among uniovular ('identical') twins of at most 25 per cent. Clearly, in this case as in others, genes are not destiny: environmental variation has an important influence. Viral infection has been suspected but not confirmed.

Similar difficulties arise with the various forms of depression. One is manic depressive psychosis: the sufferer oscillates between inertia and agitation. Mysterious changes, during a few decades, in the prevalence of certain ills, such as asthma, are well known. Depression, which may end in suicide, is also an example. During the twentieth century, in several countries, including Australia, Canada, Sweden, Switzerland and the United States, suicide, especially among young people, has risen steeply. So have other aspects of depression. Evidently, environmental causes are important; but again they have not been identified.

A reader who has followed stories in the press may now be puzzled. Discoveries of genes for mental disorders have been prominently announced, including several kinds of madness and also alcoholism (which may go with depression). They are particularly blatant examples of the 'ogod' error—one gene one defect. Later, the findings have been withdrawn by the workers who announced them; but the withdrawals have had less publicity.

Similar irresponsible reporting has encouraged belief in a gay gene supposed to cause homosexuality. The history of human biology is, as we know, spattered with examples of people throwing overboard all critical thought and of indulgence in absurdities. This is one. Two American geneticists have described abnormal behaviour by fruit flies (*Drosophila*), related to a mutant gene: the mutant males lick the genitals of other males and tend to ignore females. So we have 'gay' fruit flies. The gentle but not uncritical reader may ask: what has the sex life of flies, mutant or not, to do with human love and friendship? The answer, of course, is nothing. But the news still hit the headlines.

More recently, two more findings have had much publicity. First, a human gene for homosexuality has been reported. It seemed that the presence of a gene, whose existence had been inferred from the results of surfing the human genome, went with male homosexual conduct. Second, research has been done on the brains of a number of men, some of whom were homosexual. The hypothalamus of the homosexuals was said to be smaller than that of other males. But attempts to confirm these reports have failed.

The story is much like that concerning genes for insanity. Neither the genetics nor the neurophysiology of human sexual conduct is known. Much, however, is known about differences in custom among social groups. Both heterosexual and homosexual attitudes and practices vary greatly. According to one bizarre but influential European tradition, all sexual behaviour is sinful. Catholic doctrine (widely disregarded by Catholics) still demands that sex should be only to produce babies. In contrast, sexual intimacy, indulged in for its own sake, may be looked upon as part of the enjoyment of life. Correspondingly, in some societies homosexuality is, or has been, illegal or held to be morally wrong; but, in many others, such as the much admired classical Greece, heterosexual and homosexual conduct have been accepted as alternative ways, open to everyone, of taking pleasure with friends. Today, some communities are moving from condemnation toward Grecian tolerance. The social causes of such diversity are many and usually unknown.

HERITABILITY, INTELLIGENCE AND ENVIRONMENTALISM

Human genetics has, however, more to tell us, provided that we get it right. This is sometimes difficult, because confusing technical terms, such as heritability, are often bandied about with no explanation.

Here is a correct but misleading statement: *the heritability of human stature is high*. From this the reader might suppose that each person grows to a height determined by his or her genes. Heritability in genetics is a measure of the extent to which differences among genes contribute to variation; but it applies *only to the population and environment in which it is measured*. Environmental change, especially in nutrition (including that of the mother) and in exposure to infection, can drastically influence both growth rate and maximum height. The children of exceptionally tall parents are likely to be tall also, but only if they are brought up in conditions like those of their parents. If, in early life, they are badly nourished, exposed to cigarette smoke or seriously diseased, they may grow up stunted. Similarly, two short people are likely to have short children. But suppose the parents have been badly fed and diseased in youth, yet their children are well fed and healthy: the children may then be tall. Just this has happened, quite startlingly, during the twentieth century, in millions of families in European and other cities. I have seen the outcome at first hand in the formerly slum ridden Scottish city of Glasgow: many stunted parents with clear signs of rickets in childhood had children who towered over them.

Such environmental effects have been disconcertingly shown also by people whose parents have changed countries. Many Japanese have emigrated to the United States, where most of them are conspicuously shorter than the native Americans. But their children are several centimetres taller, on average, and their grandchildren taller still. A change of diet may be responsible.

Much more important than stature are the various kinds of intellectual and other skills. With them we come to a maze of statements and misstatements about the intelligence quotient (IQ); and we find the same interaction of nature and nurture.

The IQ has played a strange part in the dramas of education and human genetics. It began as a measure of the progress of school children in need of special help. The French originator, Alfred Binet (1857–1911), rightly assumed that teaching can improve performance at school. Eugenists, however, quickly annexed the IQ and said that it measured a fixed, inherited quality called general intelligence (or *g*).

This at first seemed convenient: here (it was supposed) was a reliable, scientific measure of scholastic worth. The result was a prolonged misunderstanding. It is still easy to get the impression that 'intelligence quotient' refers to something like the dark pigments in skin or the glucose in blood. Both can be accurately measured: they are substances. But 'IQ' does not name a substance: it is a figure derived by doing sums on test scores. A child of, say, twelve years, may be said to have an IQ of 115 and so to be well above the population average. Such a statement is based on the answers the child has given to questions chosen by examiners.

The questions usually test the kinds of knowledge taught in school or acquired at home. Among the qualities the IQ does not reveal are originality, adaptability to new ideas, critical capacity, tenacity or self restraint and the ability to plan ahead.

The IQ, in its commonest form, is therefore a number based on what a child has learned. The idea that it reflects a single intellectual quality arises from the procedure used to calculate it. If the figures are treated differently, they reveal a number of independent 'intelligences'—verbal, mathematical, spatial and others.

Findings on the IQ have another serious defect: they rarely tell us anything useful about how important abilities develop. Hence they are not a guide to those who wish to improve teaching methods. Their limitations are further exposed by a persistent muddle about the heritability of the IQ. This, though difficult to measure, is sometimes confidently said to be about 80 per cent. Such statements have been used to resist attempts to improve the teaching of disadvantaged children: poor school performance by the 'underclass' is said to be due to inferior genes and unalterable.

But, again, a high heritability does not mean *fixed by heredity*. A child's IQ can change from year to year; and, when strenuous efforts are made to better the health and opportunities of disadvantaged children, their school performance improves.

The twentieth century has indeed seen unexpected changes in the IQ of whole populations. This story begins with a myth once widely advertised in European countries, especially Britain: 'the decline in the national intelligence'. (In the United States it became 'dumbing down'.) Its origin was an argument which can be crudely stated: IQ measures intelligence; IQ is inherited; the poor are less intelligent than the rich; the poor have more babies than the rich; so genes for intelligence are leaking away and the population is becoming more stupid.

The sensible (and scientific) response was to ask whether the 'national intelligence' was indeed declining and, if so, in what sense. To answer, prolonged research was needed. This was first achieved in Scotland. In 1932, about 87 000 Scottish children aged eleven had done intelligence tests. In 1947, 71 000 did similar tests; and, contrary to prediction, they scored *higher* than the first group. Later, English children were shown to have improved too. Recently, in at least twenty countries, including the United States, the average ability to answer intelligence tests has risen substantially. The causes are not known but probably include better health.

The idea of a fixed, inherited intelligence (or *g*) is in fact an absurdity. Just as children are bigger and more healthy in some conditions than in others, so the growth of their abilities depends, to an important extent, on health, teaching and encouragement.

So now let us look soberly at the everyday tasks of parents. When a couple decide to have children, they rarely think of their genes; nor should they. A woman or a man, who wishes for healthy, happy and able offspring, should choose a partner with whom a good, lasting relationship can produce favourable conditions for rearing children. Such parents will need knowledge of a child's many needs: moral, nutritional, educational and others. The principal dangers they will face are not unwelcome genes but advertisements for unwanted junk foods and drinks, pollution, cigarettes and other hazards in the environment.

Survival and healthy development therefore depend on adopting sensible lifestyles for ourselves and our children. The recent rises in IQ have coincided with marked improvements in the conditions in which many people live. In the same period, knowing how we go about our daily lives has been repeatedly shown to be far more valuable than are revelations about human genes.

This conclusion depends on studies of large numbers of people over long periods. In one American project, the habits of 7000 people were recorded for five years. Later researches have been on a similar scale. The findings consist of statistical regularities: none can be headlined as a 'break through'; hence they do not appear prominently in the media. The importance of such studies was emphasised in 1997, in the annual report of the World Health Organization, which should have been front-page news but was not. Throughout the world, even in poor countries, people are living longer. Survival into old age is usually welcomed. Illness, however, can make aging a burden to the elderly, to their families and to society. Cancers (especially those related to smoking and bad diet), diabetes and diseases of the heart and lungs, all strongly influenced by the way people live, are rising steeply. The WHO therefore prescribes, as a matter of urgency, a worldwide campaign to encourage healthy modes of living.

To remain healthy (with a little luck) up to and during our seventies and later, we need to control body weight; take regular strenuous exercise; sleep at least seven hours a night; eat breakfast but not between meals; keep to a low fat, high fibre diet with little meat; drink alcohol only in moderation; and refrain from smoking. (And stop fussing about genes.)

To *change* to such a regime is not easy. Many, however, are attempting it. Human biology and the rules for healthy living are also beginning to be taught in schools. As these reforms take effect, the chronic degenerative diseases will decline. Elderly persons will then be able to make an increasing contribution to communal life. At the same time, good health and better schooling will enhance the abilities of children. Such statements are sometimes labelled environmentalism. They could also be called common sense.

DIFFERENCES BETWEEN GROUPS: RACE

The preceding pages are mainly about individual variation. What of differences between groups: black and white, male and female, rich and poor, slave and free? The same biological principles apply as before; but the agitated debates on these questions have less to do with authentic biology than with prejudice and politics. At different times Africans have declared themselves superior to the white, oafish inhabitants of northern Europe; Chinese and Japanese have despised people with round eyes; pale Scandinavians have treated dark southern Europeans with scorn; and in many times and places women have been assumed to be intellectually and morally below men. In this way, inferiority is forced on the supposed inferiors.

The statement that one group differs from another in a behavioural trait, even one well defined and measurable, is always likely to be misleading. Suppose that Finns could be conclusively proved to be, on average, better long distance runners than members of all other nations. (Early in the twentieth century, some people may have believed this.) The existence of Olympic gold medallists from other nations, such as Kenya, is still not ruled out. Similarly, Russians seem to be better than everyone else at chess; but other countries are producing grand masters (of both sexes) and even world champions.

To achieve excellence in athletics, chess and other skills, training is crucial. Yet we may still ask: if national or racial differences exist, to what extent are they genetically determined? The question cannot be answered: we cannot take carefully matched control and experimental groups, as though people were fruit flies or tomatoes, and subject them to rigorously regulated conditions of breeding and upbringing. Ingenious calculations are nonetheless still sometimes said to show genetically determined, average differences between people with almond eyes (the epicanthic fold) and those with round eyes; or between blacks and whites. The findings may even be headlined in the world's press.

Such averages need to be scrutinised with care. To show this, I return to stature. Here is an extreme case. The Dinka of the Sudan are among the world's tallest people; further south, the Pygmies of the forest are among the world's shortest. Only the very tallest Pygmies approach the height of the shortest Dinka. (Moreover, much of the difference is probably genetic.) Hence a committee set up to select candidates for a basketball team would not trouble with applicants from among the Pygmies.

The point of this rather absurd example is the absence of significant overlap. This is very unusual. Consider a large city with inhabitants whose parents are from South America, Europe, West Africa, India and South East Asia. Although the several groups might differ in average height, each would overlap greatly with all the others. Any of them might produce basketball players and each would also include some very short people. The differences would have both genetical and environmental causes.

Similarly, when IQ or other measures of ability are recorded, populations may differ, but a substantial overlap must again be expected. Hence persons of high ability can be found anywhere. For an example of overlap, look at the figures of stature opposite.

Averages are, however, useful, if they point to mismanagement or to wrongs which can be remedied: differences between groups may reflect disadvantage for one group. This is shown by the history of the 'intelligence' of blacks and whites in the

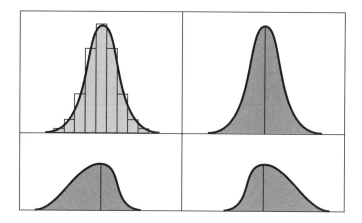

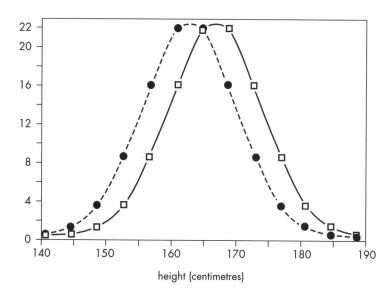

'Lies, damned lies and . . .'? Some variation within a species is simple—on or off: the reader either has blood group O or not. But most important variation is continuous. Human stature, for instance, varies continuously, from very tall to very short, and is influenced by both genes and environment. If the height is taken of every person in, say, a city, the number of persons at each height, to the nearest centimetre, can be shown in a series of rectangles, as in the histogram at top left. The smooth curve in the same figure presents the same facts.

Distributions may be skewed, as in the middle diagrams. That on the right has the form taken by the distribution of incomes—many people quite poor and a few (represented on the right) rich. When two groups are compared, great overlaps may be found. The two bottom curves illustrate overlap in stature between two human populations. All these curves show that statements concerning averages, by themselves, can be useless or misleading.

United States (the 'B/W gap'). In 1917, during the first world war, army recruits were given intelligence tests. In each State, the Negro average was lower than that of Whites; and many people at first assumed the difference to be genetically determined. But there were also large differences between States. These were so great that Blacks from some northern States scored higher than Whites in the south. Since schooling in the south was inferior to that in the north, here was evidently an environmental effect which acted adversely on children of all types. Many children are still intellectually stunted by the conditions in which they grow up. Among them are not only Blacks and Whites in slums but also Whites in rural districts of States such as Tennessee and Georgia. The same applies to many others, from the 'tribals' of India to the Aborigines of Australia.

Today, in the 1990s, we know that the B/W and other gaps can be narrowed or abolished by improving the health and education of disadvantaged groups. Yet, very recently, as an outcome of research on Asian children, history has repeated itself. Chinese and Japanese children seem to score higher in mathematics and other subjects than children of European origin; and, as usual, the difference has been attributed to their genes. It is not, however, necessary for black or white readers to get into a panic about a genetically determined 'Yellow Peril'. Asian children have also been described as more likely than Americans to enjoy school life: their schools (it is said) make an agreeable social environment with plenty of response from the teachers; Asian parents also encourage their children to do well and to attach importance to learning; and the children are happy to work hard. Hence other groups can follow suit and catch up.

SEX DIFFERENCES

The history of attitudes to sex differences resembles that concerning race. Members of the dominant group, in this case men, have until recently made all the running. Their beliefs are reflected in one of the creation stories in *Genesis*, in which Eve is a mere cutting from Adam's rib. Similarly, in some of the most influential writings of classical Greece a woman did not make the grade of full humanity. Our word, hysteria, comes from the Greek term for uterus. Hysteria was held to belong to women and women were seen as dominated by their senses and passions. Correspondingly, throughout the European middle ages women were required to be chaste, orderly, gentle and good cooks, but not clever: a woman (it was held) was sufficiently educated if she could distinguish her husband's bed from another's.

Some women, however, had always been truly learned. When, in the eighteenth century, many more were demanding education, they evoked a fierce resistance. A leading philosopher, Immanuel Kant (1724–1804), held the schooling of women to be against natural law. A century later, some men still followed Linnaeus in proposing themselves as protectors of the 'weaker sex'. They were especially horrified at the idea of women as physicians. 'No man', said one, 'could bear to imagine his beloved with a bloody knife burrowing into the bowels of a cadaver in an anatomical theatre.'

From the middle of the nineteenth century, derogation of women has continued and has had a strong biological component. In 1873, in an influential book, *Sex in Education*, Edward Clarke, a physician, stated that education for women leads to sterility and nervous breakdowns. (More than a century later, university teachers have begun to express anxiety about the exceptionally high incidence of breakdowns among male undergraduates. Women are doing better.) Stories were also told about the differences of the brains of men and women. At first, women's brains were said to be smaller than those of men; hence women were less intelligent. But women are on average lighter than men; hence their brains must be expected to be lighter too. Next, the frontal lobes were promoted as the sites of intelligence; so female frontal lobes were said to be proportionally smaller than those of men. Later, it was the turn of the parietal lobes to house intelligence; so the female parietal lobes were described as smaller. In our time, hypotheses about differences between the two cerebral hemispheres have become popular and some experts have assured us that women are left brain dominant. But the left hemisphere is supposed to be the site of verbal ability and of cold, analytical intelligence. So what becomes of those overwhelming female passions?

These stories also assume that, within the human species, brain size is an indicator of intelligence, which is not the case.

Late in the twentieth century, another biological theme has

emerged. We are, as happens so often, required to genuflect before the gene. A group of writers, the sociobiologists, try to explain all human action by the supposed effects of natural selection. It is natural, some say, for men to hunt and for women to stay at home, because this pattern has persisted from earliest times; hence it must be imposed on us by our genes. Women gatherer hunters, however, do not stay at home but are energetic foragers. The argument, in any case, does not make sense. For uncounted generations, our ancestors have worn clothes. Has clothing oneself therefore a 'genetic origin'? Was there a sudden outbreak, perhaps in the mesolithic period, of mutant genes for dressing up? (In which case, are nudists reverse mutants?) Despite the muddle, what is implied is clear: the new social roles of women in the twentieth century are against nature and therefore wrong. Whatever these writers intend, this is not biology but a social attitude.

More recently, some have resorted to new knowledge of animal courtship and reproduction. The animal kingdom provides examples of almost any sexual interaction the reader can think of—and of some no reader is likely to imagine. A male salamander (*Amblystoma*) sometimes intrudes on a mating pair and imitates the female. The courting male is provoked to release his sperms and these are promptly covered by the intruder's sperms. In this way the intruder fathers the female's young without having to trouble about courtship. I suggested, in the scientific journal, *Nature*, that such conduct should be called caddish behaviour. This phrase has, regrettably, not been adopted as a technical term; but now more such conduct has been found, among animals from dragonflies (Odonata) to small birds called dunnocks or hedge sparrows (*Prunella modularis*).

The critical reader may now ask, as before: what have salamanders, dragonflies and dunnocks to do with our present subject? The answer, again, is nothing. Yet these and other animals are used by modern storytellers to support their pictures of human conduct. According to some, a woman is programmed by her genes and hormones to produce as many healthy, fertile young as possible. To achieve this, she should have children by several selected fathers and in this way increase the chances that her genes will survive. We are even told that, at a certain stage in the monthly cycle, women are under a compulsion to be unfaithful to their husbands.

Here is not only a weird fantasy: it is a harking back to the tradition of women at the mercy of their passions. Such inventions not only make no allowance for love or friendship but also neglect a commonplace fact: many women now decide, after careful thought and discussion with a partner, just when they will have a child and by whom. This is one aspect of the current rapid and, to some, upsetting changes in social roles and attitudes.

The changes are not confined to marriage customs: they are also reflected in job allocation. Until recently, clerical and similar duties were performed by men. The employment of women as clerks was at first angrily resisted. Yet now, in many countries, most of this work is done by women. More disconcerting to some people are the women employed as physicians, pilots, priests, prime ministers and presidents. In the United States, in 1970, women made 16 per cent of managers, executives and administrators; 25 years later, the figure was 43 per cent. Such changes illustrate the principle that, when we look for ability in two populations (in this case, women and men), we must expect to find, in each, persons with the required qualifications.

The attitudes of men toward women, expressed in the incoherent writings of several millennia, reflect principally conventional ignorance and prejudice. No doubt, men have often felt love, affection and even admiration for their wives, mistresses and daughters—but, as a rule, only from a position of superiority.

*

To sum up, modern genetics has not enabled us to identify genes for moral worth, intelligence or special skills. Nor have genes for poverty, general ill health, insanity or criminality been found. These are not simple traits: a person is not worthy or unworthy, bright or stupid, healthy or unhealthy, sane or insane, heterosexual or homosexual, like flowers coloured either red or white. Genetical variation certainly does contribute to our differences; but, when we describe a person as healthy or good or intelligent or skilful or sane, we are referring to complex attributes of a human being: each feature has many components and its development is influenced by a succession of diverse and changing environments.

Abilities and other desirable qualities can emerge in any human community, regardless of physical type, class or sex. They cannot be reliably predicted, only encouraged or discouraged. The many kinds of skill and virtue are distributed erratically and may appear without warning, especially when people have access to education and training. If skills are to be fully developed, children of all kinds and both sexes, regardless of parentage, must have good health and access to education in a wide range of subjects.

More generally, for human development *some environments are much better than others*; and, to some extent, *we can make or choose our environments and those of our children*. Recognition of these facts has already allowed impressive progress, in many countries, in standards of health and education. Whatever genetic variation exists, to continue this advance we need to concentrate on the conditions in which we and our children live.

Chapter 5

The Smallest Organisms: Matters of Life and Death

> See where Amoeba marks the printed page!
> Her formless body tells the attentive sage
> Of how our forbears lived in ancient seas.
> Others, more sexy, change by slow degrees
> To larger creatures. These, with many cells,
> Develop nerves. And then a ganglion swells,
> Becomes a brain — and look!
> 'Amoeba has her picture in a book'.
>
> ANON (Last line from Julian Huxley)

THE CONDITIONS IN WHICH we live are continually influenced by beings which were, until the seventeenth century, completely unknown. We therefore now return to the story of microorganisms, begun in chapter 3. They are a very mixed bag and so is this chapter. The simplest are the prokaryotes, single cells with DNA but no nucleus, which are the nearest we have to the first forms of life. They surround us; we depend on them for food and much else. But some kill us.

In favourable conditions a bacterium can divide every twenty minutes. After eleven hours, the outcome would be 5000 million descendants of an initial single cell, a number similar to the human population of the world in 1990. As a result, bacteria are everywhere in the biosphere in vast numbers. Their importance has been recognised since mid nineteenth century, yet they can still surprise us.

Our commonest pathogenic bacterium was discovered only in 1979. It lives in one of the least favourable environments offered by our bodies, the stomach. *Helicobacter pylori* thrives in acid gastric secretions which kill most organisms. A hairlike flagellum propels it in the sticky fluid on the inner surface of the stomach. It also produces ammonia which neutralises the acid around it.

H. pylori was discovered in patients with inflamed stomachs. One of the discoverers, an Australian, Barry Marshall, whose stomach was uninfected and healthy, swallowed some of the bacteria and soon developed a severe and painful gastritis. (He recovered quickly after treatment.) Further research showed the several forms of the bacterium to be a major cause of gastritis and of gastric and duodenal ulcers. Gastric ulcers had been regarded as due to excess acid induced by 'stress'. They had been treated with prolonged courses of antacids and other drugs, sometimes by surgery or even psychotherapy. Today, gastritis and both kinds of ulcer can usually be cleared up in a week or two by an inexpensive regime of antibiotics.

This happy ending is, however, only a beginning. It is not known how the infection is acquired. Often, it seems, it begins in childhood and causes illness only after many years. Have other slowly acting pathogens still to be found? Probably, yes.

Bacteria Help Us

As *Helicobacter* shows, bacteria are chemically very diverse. Some can 'fix' the uncombined nitrogen in the air or in water. Nitrogen fixing bacteria on the roots of leguminous plants (lentils, beans, peas and clover) famously help to restore soil fertility after other crops have used it up. The nitrogen fixers balance the effects of denitrifying microorganisms which break down nitrogen compounds and release nitrogen into the air or water. This is the nitrogen cycle, a major part of the continuous recycling of elements in the biosphere.

Among nitrogen fixers are simple green organisms, the cyanobacteria. The stromatolites of 3500 million years ago contain traces of these 'blue-green algae'. All known species can carry out photosynthesis, as well as fix nitrogen. They are numerous in all waters and in soil. In Antarctica they inhabit both icy lakes and hot springs. They make a large contribution to a conspicuous feature of our planet, its green colour, seen brilliantly in photographs taken from space; but some are reddish and are responsible for the occasional strange appearance of the Red Sea. Cyanobacteria can live on carbon dioxide and nitrogen from the air, together with simple inorganic substances dissolved in water. In 1883, an island, Krakatau, in the East

Indies (Indonesia), was blown up by volcanic action. All life was destroyed. Among the first organisms to reappear on the shattered surface were blue-greens, borne as spores by the wind. They helped to prepare the way for the return of larger organisms.

Other microbes represent for us, more immediately, both life and death. The sea, fresh waters and the soil are seething with the organisms of decay. A teaspoonful of fertile soil (say, 7 grams) contains at least 5000 million bacteria (as well as smaller numbers of yeasts, moulds and amebae). All larger organisms both depend on them and are eventually destroyed by them. Putrefaction, usually disagreeable at close quarters, is the breakdown of dead organic material, largely by bacterial action, and its conversion to simpler substances. It is essential for soil formation. In its absence, dead matter would pile up and choke the biosphere, like a city smothered in refuse when the garbage workers go on strike. Other kinds of putrefaction are welcome for other reasons: they give us cheese and yoghurt.

Sheltering in our intestines, certain bacteria synthesise vitamin B_1 (thiamin), an essential component of our diet. When the reader takes an antibiotic to treat an infection, these helpful organisms may be destroyed. The remedy is to reinfect the gut by eating a special yoghurt. Such symbionts are widespread. Ruminants, such as cattle, depend on microorganisms in their stomachs. Horses have similar bacterial populations in a branch of the gut, the caecum.

Some bacteria thrive in seemingly impossible conditions—worse even than stomach acid. Among them are salt lakes thick with sodium chloride; freezing sea water; strongly alkaline waters; rocks deep under the sea; and hot, acid springs where the water perhaps slightly resembles the thin soup in which organisms may have first evolved. All these have enzymes which perform remarkable feats. As a result of 'engineering' their genes, we can now use bacteria to carry out tasks which formerly put heavy demands on chemists or engineers. Some separate gold and copper from mineral ores. Others remove the pollution caused by disastrous oil spills from tankers wrecked on fertile shores. On a smaller scale, yet others deal with residues, mainly benzene, from old petrol stations. According to some enthusiasts, microorganisms exist which can metabolise anything.

Amoeba proteus, aka *Chaos chaos*: famous; sexless; shapeless?, primitive or degenerate? Whatever, a very successful organism. This one is showing its prowess as a predator: the prey is a smaller protist. Cells like this, in body fluids, ingest invading microorganisms.

The Protista: Single Cells with Nuclei

Many microorganisms have their DNA enclosed in a nucleus. These eukaryotes include the protozoans, of which *Amoeba proteus* has long been among the most popular. It once seemed a perfect example of a 'primitive' animal: one cell; no shape; no sex; reproduction by a mere splitting: even the nucleus seemed to undergo an informal fission into two. At one time it was given the Latin name, *Chaos chaos*.

We no longer see *Amoeba* like that. It is not shapeless. The flowing ameboid progress, by pseudopodia or false feet, is a mechanical marvel. During movement, food particles are identified, engulfed and eaten, as if this one cell possessed a nervous system and sense organs. The DNA in the nucleus is organised in chromosomes but these are many and, even under the highest power of an ordinary microscope, look like dust grains. The absence of sexual reproduction is held to be not 'primitive' but 'degenerate', which signifies only that we assume ancestral amebae to have reproduced sexually, as other members of its group do today.

Other protistan eukaryotes, like 'blue-greens', have chlorophyll. Green, single-celled organisms with nuclei flourish in multibillions in the sea and in lakes and rivers. Among them

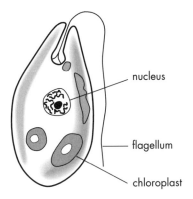

Plant or not plant—or neither? The ambiguous flagellate protistan, *Euglena viridis*—a single cell which can be green or not green.

the diatoms are each lightly armoured in a silica box perforated by many holes. Despite their microscopic size, during photosynthesis they use more carbon dioxide and manufacture more organic matter than do any other organisms. They also start off a food chain which leads through small animals, through larger ones such as fish, to still larger ones, such as people.

Some green cells move by means of flagella. The most disconcerting are sometimes green, sometimes not, like politicians trying to attract votes from both environmentalists and loggers. *Euglena* is easily found in freshwater pools and even in mud, where it forms a green scum. Transferred to a dark cupboard and kept in a suitably mucky solution, it loses its chlorophyll and lives on organic matter. But, returned to the light, it becomes green again; it then moves, like an animal, to wherever the light is strongest. Hence *E. viridis* appears in textbooks both as an animal and as a plant.

THE SCOURGE OF MALARIA

Some protozoans live in large animals. The malarial parasites (genus *Plasmodium*) infect monkeys, chimpanzees and gorillas and seem to do them little harm. But, for millennia, people in large areas of the tropics and subtropics have been debilitated by occasional bouts of fever or, sometimes, by continuous illness. And still, today, of all human infections worldwide, malaria causes the most illness, misery and impoverishment. Malaria can kill; but more often it causes a chronic weakening. Figures collected in the 1980s indicate, for the whole world, 200 million cases a year and two million deaths. Epidemics often occur in spring and autumn, the seasons of sowing and harvesting, and may have a calamitous effect on a farming community. Its effects (with those of other widespread diseases) have encouraged the belief, among uninfected Europeans, that 'natives' are naturally idle.

The name, malaria, comes from associating the disease with low-lying, stinking wetlands. The illness, however, is due to protozoans which intermittently become numerous in the blood and destroy many of the red blood cells. The very small parasites have a complicated life story and the long struggle of discovery, treatment and prevention was complicated too. It began with Alphonse Laveran (1845–1922), a French army surgeon, who identified the parasites in 1880 while posted to Algeria. In 1898, in Bengal, Ronald Ross (1857–1932), a Scottish poet and medical officer in the British Indian Army, demonstrated their transmission by bloodsucking mosquitos (genus *Anopheles*) which breed in marshes, ponds and streams. When an infected mammal or bird is bitten, *Plasmodium* enters the mosquito's stomach with the sucked in blood and forms minute swellings in the stomach wall. Later, the parasite moves to the mosquito's salivary glands, from which it can be injected into another host. This major discovery was made by studying infected birds.

Soon afterwards, an Italian physician and zoologist, G.B. Grassi (1854–1925), carried out a series of well designed, ruthless experiments on large numbers of human beings. In parts of the south of Italy malaria was a serious problem. Grassi showed that, in such areas, people protected from mosquito bites did not become infected. Others were left to be bitten and many succumbed.

Even then, an enigma remained. A feature of malaria is its recurrence long after apparent recovery and after return to a country where reinfection can be ruled out. It was only in the 1940s that *Plasmodium* was shown to lurk in the liver: from there it can reinfect a patient long after the first bout of fever.

In mid century, the World Health Organization, armed with much accumulated knowledge, began a grandiose program to eradicate malaria: the mosquitos carrying the parasites were to be wiped out by insecticides, especially DDT. But, after vast expenditure and effort, the project failed. The adaptability of organisms had not been allowed for. The mosquitos developed resistance to the insecticides. Their behaviour also changed and they became less accessible to sprays. The malarial parasites have changed too: many populations have become resistant to chloroquine, the antimalarial drug recently most used.

By the 1990s, among the two-thirds of humanity living in malarious regions, the incidence of the disease was again rising and it has once more become a menace to travellers. New drugs are urgently needed. But, in the words of an American authority, Robert Desowitz, 'No major efforts have been made to find an antimalarial to replace chloroquine. The motivations of colonialism and profit are gone.' The latest news, however, is of a vaccine which will perhaps protect those who can afford it.

THE YEASTS AND MOULDS: WINE AND CHEESE

After that, we need cheering up. Other organisms invisibly influence our lives for the better, notably the yeasts. These, though single cells, are usually classed with the fungi. Each cell has a nucleus with chromosomes and, in the cytoplasm, a complete set of organelles and often an oily food reserve. A few form strings of cells: they then look more like typical fungi.

The most famous is baker's or brewer's yeast (*Saccharomyces cerevisiae*). Put a few cells in a solution of glucose, and the sugar is combined with oxygen; carbon dioxide and water are released and energy is produced (aerobic respiration). Meanwhile, the yeasts multiply rapidly by budding—that is, asexually. For this to happen, the solution must be well aerated. Without oxygen, fermentation occurs: the yeasts convert glucose to alcohol (ethanol) and carbon dioxide, again with the release of energy (anaerobic respiration).

Wine and other beverages are produced by this process. Until Pasteur's famous experiments, fermentation, like decay, was supposed to occur 'spontaneously'. Pasteur showed, with typical skill, that grape juice would not ferment by itself but would do so if washings from the skin of grapes were added. Washings did not have this effect if they were first boiled. Boiling killed the yeasts present on the skin of mature grapes. Hence to make wine, infection of the grape juice ('must') with yeasts is needed.

When we talk of wine, we usually mean a liquor derived

from the fruits of the vine (*Vitis vinifera*), of which many varieties are grown in six continents. But other fruits, such as damson plums, may be used in the home. Not even fruits are essential. Some people think highly of wine made from the roots of parsnips (*Pastinaca sativa*), which contain much sugar as well as starch. In the tropics many kinds of palm wine are made. The juices of the coconut palm (*Cocos nucifera*) and the date palm (*Phoenix dactylifera*) are very sugary. In each case, the material, like grapes, must first be crushed; but trampling with bare feet is not essential.

Producing wines in the home can make an interesting experiment. So can drinking them. Most wines are bottled only when carbon dioxide production has stopped. Sometimes, too much carbon dioxide is produced by a home brew and the container explodes. The bubbly wines, of which the most famous come from the French province of Champagne, presumably originated by error, when wines were bottled too early.

MOULDS

Change and decay in all around I see.

The moulds are fungi which can grow in almost any environment which contains dead organic matter. (This includes the timber in houses: dry rot, which can wreck a building, is due to a fungus which lives on dead wood.) Together with bacteria and yeasts, they are responsible for the essential processes of decomposition: dead organic matter is broken down; carbon dioxide is released into the air; nitrogen compounds and other substances are excreted and recycled in the soil. The top 20 centimetres of one hectare of good soil may contain five tonnes of fungi and bacteria.

Mouldy bread and meat are regarded as unfit for human consumption, except by the very hungry. Mouldy cheese, on the other hand, can be a delicacy. The delicious green and blue cheeses, such as Stilton (England), Gorgonzola (Italy) and Roquefort (France), owe their character as well as their colour to moulds. No doubt, like champagne, they originated by accident. A piece of quite uninteresting mild cheese is said to have been left by a peasant boy in a cave. When he returned some weeks later, he found it mouldy and much improved. The fungus responsible, *Penicillium roqueforti*, was first identified in caves near the village of Roquefort.

For creation of the best sweet white wines of France, a mould is essential. In the communes of Sauternes and Barsac, some grapes are left to rot on the vines. The decay is due to *Botrytis cinerea*. The result is *pourriture noble* or noble rot—a stage in the production of *Chateau d'Yquem* and other such famous wines. They are best when drunk, in small amounts, at the end of a meal.

Fungal infections of fruit trees are, however, usually unwelcome. In the 1850s, French wine production was brought near to ruin by an outbreak of mildews which killed young vines. By 1854, production of wines and spirits had been reduced by 80 per cent. American vines which resisted the mildew were extensively planted, but they brought with them an aphid, *Vitis vitifolii*. This insect, the vine phylloxera, attacked the remaining French vines; and, worse, another mildew, *Plasmopara viticola*, arrived too. The decline in French wines and their products in mid century led the English to turn from French brandy to what they had previously regarded merely as foul fire water, namely, Scotch whisky.

The remedies, however, were quickly applied: American rootstocks resisted the insects; French cuttings, grafted on to the rootstocks, produced the best grapes; and chemical treatment (with Bordeaux mixture) attacked *Plasmopara*. The result was a marvellous recovery. Mouton Rothschild 1899 (a top claret) was still superb in 1939.

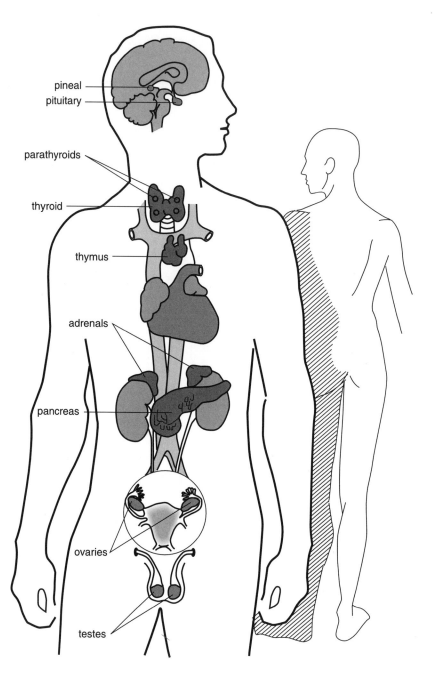

'Where the hormones, there moan I'? A hermaphrodite drawing with some of the principal endocrine glands of human beings. For their functions, see pages 102–3.

The Importance of Being Constant: Bodies and Minds

It is sometimes useful to look at an organism, even a human being, as a self-maintaining mechanism. Regulation, however, often depends on inconstancy, especially in behaviour. More important, a person is not a machine. People have minds of their own.

Chapter 6

The Stability of the Body

> If this be magic, let it be an art
> Lawful as eating.
>
> SHAKESPEARE

An experimenter once arranged to spend several days shut up in a laboratory in which all his bodily exchanges of matter and energy could be audited. The amounts he ingested of protein, fat, carbohydrate and water were recorded; his urine and feces were collected and analysed; losses through sweating were estimated; and his output of carbon dioxide was measured. As he expected, input and output balanced with as much exactness as the methods of measurement allowed. Correspondingly, his weight was the same at the end of the experiment as at the beginning. Since then, others have been subjected to similar conditions, with similar results.

Such findings encourage us to look at human and animal bodies as chemical machines. A pioneer of this method was Claude Bernard (1813–1878), the son of a French peasant. As a result, Bernard, like his still more famous contemporary, Louis Pasteur, was involved in controversy with colleagues. Many of them held that life is due to, or consists of, a vital essence outside the scope of physical science. Bernard rejected this view. He was an experimenter of great skill who showed how much can be learned about the body by concentrating on what can be directly observed and measured. His *Introduction to the Study of Experimental Medicine* became a foundation of modern clinical methods. In a healthy person, temperature, blood pressure, blood sugar level and many other measures are remarkably constant; the tissues are therefore kept in favourable surroundings and at a steady temperature.

The maintenance of an almost constant internal state we now call homeostasis, a word owed to an American physiologist, W.B. Cannon (1871–1945). Not all Cannon's work was about steady states. When an emergency demands violent exertion, survival depends on large changes in some organs. Cannon made a leading contribution to understanding them in his *Bodily Changes in Pain, Hunger, Fear and Rage*, appropriately published in 1915, during the first world war; and he began the study of the adrenal glands during 'stress' (another term he brought into physiology). Later, as an army medical officer, Cannon saw the effects of horrific wounding at first hand. He also experienced severe stress himself.

Today, the greatly raised secretion of a hormone, adrenalin (epinephrine), by the adrenals, when a person is frightened, furious or in pain, is part of common knowledge. This, with other hormones produced by the adrenals, helps to keep the body going in adverse conditions. In an emergency, adrenal hormones stimulate conversion of liver glycogen to sugar which, passed into the blood, provides the muscles with extra energy.

Other discoveries concerning hormones, early in the twentieth century, made it possible to treat previously baffling diseases: people with some kinds of diabetes mellitus, formerly fatal, could be kept in good health with injections of insulin; disorders of the thyroid, the pituitary and other glands could be relieved. Later came even more important applications, especially the contraceptive pill.

Certain hormones, including the thyroid hormone and those concerned with reproduction, influence mood and behaviour: hence zealots began to talk of solving the problems of mental illness and criminality by endocrine treatment. By the 1920s, these ideas had already appeared in popular novels. A successful detective story, *The Unpleasantness at the Bellona Club* by Dorothy Sayers, has a physician who is said to hope to 'make everyone good by glands'. A reference is also made to 'casting out devils with a syringe'. (The doctor in the story turns out to be the villain.) It is always tempting to hope for simple solutions to bewildering problems. Today, however, nobody seems to regard hormones as cure-alls. Other panaceas, especially genes, have become fashionable. More of them later.

Keeping Warm and Keeping Cool

The chemical processes in our cells need a temperature inside the body at or near 37 degrees Celsius. If the blood reaching the brain cools, nerve impulses pass from the hypothalamus, at the base of the brain, to the rest of the body. Muscles contract in the walls of small arteries which carry blood to the skin. Less blood reaches the body surface and loss of heat is reduced. Another rapid response is shivering, an involuntary action which produces much extra heat. Its effectiveness is seen in an extreme form in some infections. An early sign of illness may be a fit of violent shivering (a rigor) which can quickly produce a raised temperature. This, of course, is a disturbance of the normal state; but the fever perhaps accelerates production of antibodies and so helps recovery. Continued exposure to cold requires a general rise in heat production by the body's tissues, with no extra muscular movement, which allows a naked human being, accustomed to exposure, to survive in temperatures below zero.

Despite insulation by skin and fat, heat is continuously lost from the body surface. Suppose you were put in a giant vacuum flask which allowed breathing but no loss of heat. Even if you kept perfectly still, your temperature would rise by more than 1°C every hour. Death would occur after about five hours. Occasionally, just this sort of disaster actually happens, for instance if a small baby is left in a closed car in the sun.

In heat, as in cold, receptors in the brain are activated; but now the blood supply to the skin increases, the skin reddens and more heat is lost. Still more important is sweating. Almost every square centimetre of skin has several hundred sweat glands which open on the surface. When we get hot, large amounts of watery sweat are produced and its evaporation cools us. The cooling efficiency of sweat is prodigious. Physiologists indifferent to discomfort have shown this by entering a large oven heated to a cooking temperature, and staying long enough for a steak to be well done, say, 45 minutes. One took a dog with him in a basket. The dog survived by panting and secreting watery saliva on to its projecting tongue—the canine equivalent of sweating. (Afterwards, it also ate the steak.) The air in an oven is dry and allows sweat, or a dog's saliva, to evaporate quickly. In saturated air, in which sweat could not evaporate, one would not survive such temperatures for long.

Some human endocrine organs and hormones

Pituitary gland: has two main parts and secretes many hormones, some of which act on other endocrine glands. Has been called 'conductor of the endocrine orchestra'.
'Anterior' pituitary: Growth hormone; gonad-stimulating hormones; thyroid-stimulating hormone; adrenal-stimulating hormone (acts on adrenal cortex).
'Posterior' pituitary: Oxytocin causes uterine contraction; may be given during childbirth to help in expulsion of the infant.
Vasopressin causes kidney to retain water; deficiency leads to excessive production of urine (diabetes insipidus) and severe thirst: treatable by giving the hormone.

Thyroid gland: the thyroid hormones, which contain iodine, stimulate the breakdown of sugar in cells. Deficiency causes goitre, sluggishness and intellectual impairment (cretinism): treatable with thyroxin. Excess (thyrotoxicosis) causes extra activity, agitation and weight loss: treatable by anti-thyroid drugs or removal of part of thyroid.

Pancreas: Insulin stimulates cells to take up sugar (glucose) and amino acids from the blood. Deficiency causes diabetes mellitus with excess of sugar in blood and urine, wasting and possible death: treatable by diet or by giving insulin. Glucagon has an effect opposite to that of insulin: it stimulates the release of sugar into the blood, where it can reach the tissues, and is important during starvation.
(The pancreas also secretes enzymes which take part in digestion in the small intestine—an *exocrine* function.)

Adrenal glands have two parts.
Medulla: Adrenalin (epinephrine) and noradrenalin (norepinephrine) are quickly secreted in large extra amounts when sudden vigorous activity is needed; they increase the amount of sugar available in the blood.
Cortex: Secretes several adrenocortical hormones (glucocorticoids). Main effects include stimulating synthesis of glucose from fats and protein; hence operate in emergencies, but less quickly than hormones of medulla. In conditions such as continued exposure to cold, adrenal cortex enlarges.

The gonads, as well as producing eggs or sperms, secrete hormones.
Ovaries: One ovarian hormone, estrogen, has many effects. At puberty it stimulates growth generally and enlargement of the mammary glands (breasts). Progesterone, another hormone, and estrogen together regulate the estrus ('heat') cycle of most mammals and the menstrual cycle of human beings. Estrogen also acts on the brain and influences behaviour. Substances similar to these ovarian hormones are used to treat disorders of menstruation and for birth control ('the pill').
Testes: Testosterone, the male hormone, too has many effects. At puberty it stimulates growth generally, enlargement of the penis, growth of the beard and production of sperm. It also influences behaviour.

The list above, of glands, of hormones and of their effects, is far from complete.

KEEPING FED

Even in a constant, favourable environment, homeostasis requires continual exertions by the body. Oxygen is the most urgent necessity. For most mammals, the next is water.

After that comes food. A regular intake depends on the brain. One small region of the hypothalamus is especially concerned with beginning a meal; another, with stopping it. An animal with the first one out of action refuses all food. If the other is injured, the animal overeats and becomes obese: in a few weeks a rat weighing 200 grams may grow to 400. A regulatory device (a 'ponderostat') has evidently been reset to a new value. In a normal animal or person, the hypothalamus responds to neural and hormonal signals from the organs which process and store food. The more these processes are studied, the more complex the many feedback loops (illustrated on page 39) are found to be. A recent unexpected finding is of hormones secreted by adipose tissue. One of them, leptin, reduces appetite.

We cannot, however, rely wholly on the automatic physiology of feeding. Many readers of this book are likely to be in danger of eating too much: the appetite of some people is set at a point which allows the slow growth of adipose tissue. In a rich country, such as the United States, after the age of twenty-five a person typically gains about 20 kilograms in 30 years.

Obesity increases illness such as heart disease and reduces expectation of life. How can it be prevented? To get fat, an adult must eat more food than is used for movement, for producing heat and for tissue repair. Correspondingly, if *less* is eaten, existing fat stores are reduced and weight goes down. So, it may seem, to lose unwanted fatty tissue should be easy: exercise more and eat less.

But such an alteration of habit is hard to achieve. It is therefore often said that some of us are genetically forced into fatness. Publicity has also been given to 'genes for obesity' discovered in mice, but their significance has been exaggerated. Although people doubtless differ genetically in their capacity to become obese, whether one is obese or not depends greatly on what one eats. This has been shown by the outcome of the changed diets of whole populations. The inhabitants of a Pacific island, Nauru, have recently become rich by selling guano for fertiliser. Formerly, they exerted themselves as farmers and ate vegetables and sea food. Body weight was not a problem. Now, they have adopted Western diets and lifestyles. As a result, about two-thirds are obese and one-third are diabetic.

Here we again meet the fundamental principle of epigenesis: a trait, such as obesity, reflects the influence of both the genes and the environment. And, as usual, it is on environmental effects, and on personal will, that we have to rely for the most effective action. One recent unsurprising discovery concerns the adverse effect of dietary fat: to prevent obesity or to lose weight, one should cut down on fried foods, butter, cream and ice-cream. An austere diet, combined with regular, vigorous exercise, is the best policy. This does not, however, mean that one should give up fats altogether: fats are an essential component of our diet.

TOIL AND SWEAT

Exercise should not be confused with sweating. Sweat contains salts and other substances but is mostly water. The water is quickly replaced by drinking. If it is not, in addition to severe thirst, kidney damage is likely. Slimming, as jockeys and fashion models know, cannot safely be achieved by dehydration.

Omnivores, such as human beings, must not only eat enough: they must also choose appropriately from whatever foods are available. Many mammals, given a choice of foods, have some capacity to select the right one. Suppose rats are fed for a time on a diet with little thiamin (vitamin B_1) and are then put in a 'cafeteria' situation with several food mixtures. They sample all; but they soon learn to eat mainly the one with plenty of thiamin. Mammals respond similarly to other dietary essentials, such as salt. So do human beings. Men, such as miners and stokers in old-fashioned steamships, who do hard work in hot conditions, lose much salt in sweat: they make it up by eating salt food and even by drinking beer to which salt has been added (very disagreeable if one is not salt deficient).

More surprising, young children have been allowed to choose freely among a large number of foods. Over a period, they spontaneously achieved a balanced intake. In that experiment, all the foods offered were capable of contributing to a healthy diet. Parents should not offer their children starchy or sugary foods or sweet drinks which are enjoyed (and ingeniously advertised) but are liable to lead to malnutrition. Certain breakfast cereals with a high proportion of sugar are presented as providing energy and so they do; but health depends on eating foods, such as whole grain cereals, which also contain plenty of fibre and adequate vitamins (especially those of the B group) and little sugar.

The obverse of choosing favourable foods is avoiding unfavourable ones. Omnivorous mammals test all the foods available. If they sample a toxic mixture and become ill, they stop eating. When they recover, they refuse the poisoned food or anything that tastes like it. (That is one reason why it is difficult to get rid of wild rats by poison baiting.) Human beings have a similar ability. A reader who becomes ill shortly after eating a particular food may refuse that food for years afterwards.

All this knowledge, though valuable, hardly helps the millions who are continuously hungry. In 1992, the Food and Agriculture Organization estimated the world's severely undernourished at 800 million. Of them, in the 1990s, an estimated 200 million children are growing up stunted and with low resistance to disease. Twelve million are in the United States, the world's richest country. These children are both physically dwarfed and intellectually impaired. Careful and intensive studies of poor families in several countries, including Guatemala, Jamaica, Peru and the USA, have all pointed the same way: the ability of a child to benefit from school teaching is greatly influenced by what he or she eats. Village children and pregnant women have been given a palatable food supplement with much protein and also vitamins and inorganic salts. Later, the children were tested for language and arithmetical skills and general knowledge. They did better than similar children, in other villages, who had received a less nourishing supplement. This work is of special importance for reasons given in chapter 4. Some influential persons, fixated on genes, strenuously deny that remedial measures can improve intelligence or cognitive skills.

Human nutrition: the main types of foodstuff

Needed in large amounts: sources of energy

Protein	Eggs, milk, cheese, legumes, meat, fish.	Especially important for growth.
Carbohydrate	Cereals, potatoes, yams, sugar.	Main source of energy for most people.
Fat	Butter, cheese, some meat, some fish.	Twice the energy value of carbohydrate.

Needed in smaller amounts: inorganic substances

Salt	Most foods.	In all body fluids.
Calcium & phosphorus	Milk, cheese, vegetables.	In all tissues; a large part of bones and teeth.
Iron	Vegetables, fruit, whole wheat, red meat.	Contained in hemoglobin.*
Iodine	Vegetables, eggs, meat.	Contained in thyroid hormone.*

Needed in very small amounts: vitamins

A	Fish liver, milk, butter, eggs, green vegetables.✢
B group	Whole cereals, meat, eggs.*
C (ascorbic acid)	Citrus fruits, tomatoes, other fresh fruits and vegetables, fresh meat.○
D	Fish liver, milk, butter, eggs, (sun on skin).✢

Vitamins A and D are fat soluble. Those of the B group and C are water soluble and easily destroyed in cooking. (The list of vitamins is incomplete; but those not mentioned are of less everyday importance.)

Some deficiency diseases

✣ Iron deficiency causes weakness due to anemia: the amount of hemoglobin, the red pigment of blood, in the red cells is low. Anemia interferes with intellectual development in children. It has been described as the world's commonest dietary deficiency. Women with a heavy menstrual loss of blood are especially at risk; also people who lose blood as a result of parasitic infections.
✣ Iodine deficiency can cause goitre and cretinism. Most likely in mountainous regions where the iodine content of the soil and of plants is low. Now usually prevented by adding small amounts of iodide to table salt. (See thyroid, p. 102.)
✦ Vitamin A deficiency can contribute to skin disorders such as acne. Severe deficiency leads to eye disorder (xerophthalmia) and even blindness.
★ The B group contains several vitamins. B_1 (thiamin) is needed for normal function of the nervous system. Minor deficiency can cause chronic fatigue. Severe deficiency causes beri-beri, a potentially fatal condition of extreme weakness with muscular spasms. B_2 (riboflavin) has many functions. One effect of minor deficiency is inflammation of the lips and mouth; major lack causes degeneration of the cornea of the eye. Minor deficiency of B_3 (niacin or nicotinamide) causes diarrhea; severe deficiency causes pellagra: the skin is scaly and sores develop; can be fatal.
✺ Minor deficiency of vitamin C causes anemia and slow healing of wounds; severe lack leads to a weakening and sometimes fatal disease, scurvy, once common in late winter in European towns; a notorious hazard in the past for sailors on long voyages.
✣ Vitamin D, which includes several related substances, is needed for normal metabolism of calcium and therefore for growth and maintenance of bones and teeth. In the recent past, many children in northern cities developed rickets, a distortion of the skeleton, owing to deficiency. Vitamin D is formed in the skin in sunlight; hence rickets is rare in sunny climates.

Vitamin deficiency is unlikely among people who eat a balanced diet with plenty of fresh food, whole grain cereals and no excess of sugar or alcohol.

Is 'Natural' Food Best?

The human species has existed for perhaps 100 000 years. During nearly all that time, food was acquired by gathering plants, catching or trapping animals and scavenging for dead or dying meat. Such an existence is sometimes said to be natural for human beings. In the same way, wild horses may be said to be living natural lives, in contrast with those kept in fields and bred in captivity. The word 'natural' then refers to the conditions in which the species has evolved.

We may therefore speak of natural human food, meaning what people ate before agriculture was invented more than 10 000 years ago. Such people are called 'hunter gatherers' but 'gatherer hunters' is more appropriate. A few remain, in African deserts and forests, Australian deserts and elsewhere. Before exposure to Europeans, the Aboriginal Australians seem to have eaten a great variety of foods, including small animals. Their diet was low in fat, high in fibre. In body build they were lean but they were not chronically hungry. Ills common among Europeans, such as diabetes, cardiovascular disorders and obesity, developed only when European habits were adopted. If Aborigines return to their traditional diet, their health quickly improves.

These findings correspond to the experience of the islanders mentioned above. They also match the advice of nutritionists on how we should live. We need a diet based largely on whole grain cereals, vegetables and fruit, with eggs and cheese but little fat; and perhaps some meat and fish. And we should take regular, enjoyable, strenuous exercise.

This advice does not, however, arise from knowledge of people living in deserts or forests without agriculture. The very few gatherer hunters excepted, human beings depend on the food stores of cultivated plants which have been greatly altered from the wild types. When we eat potatoes, pasta, tapioca, porridge, bread or rice pudding, we use starch which has been stored in seeds or tubers. Plants also store protein in their seeds: those of legumes (lentils, dahl, peas and beans) have more than usual; they therefore have a special value in the diet of vegetarians. Some plant structures, such as nuts and olives, have high concentrations of fats or oils.

The main civilisations have been founded on the cultivation of selected rice (in Asia), potatoes (in Central and South America) and wheat and barley (in the Middle East and Europe). The staple food plant depends on the climate, hence on what is available: banana republics are confined to regions where the banana palm can grow.

Current recommendations on diet come from researches on

people, such as the reader, who live in cities or other modern environments and depend on increasingly modified food plants. A careful selection of the available foods (most of which are very 'artificial') can yield a healthy diet. The important question is not whether they are 'natural' but whether they are good for health.

FOOD FOR THOUGHT

Chapter 3 describes something of what happens to foodstuffs when they enter cells. The account there is at the level of cells and their chemistry. Much of the present chapter is concerned with interactions between organs of the body, such as adipose tissue and the brain: this is the level of physiology. Behaviour also comes into the picture: this is the domain of ethology or of physiological psychology. To understand how an animal or a human being is kept going, we need to know what happens at all these levels. In this way we approach a complete picture of a body which can be studied as a physical system—the method advocated by Claude Bernard.

But some of our dietary habits have little to do with homeostasis. They illustrate the influence, always present in human communities, of 'culture'. Conventional diets reflect local custom, exemplified by a notorious entry in Samuel Johnson's *Dictionary of the English Language*: 'Oats. A grain, which in England is generally given to horses, but in Scotland supports the people.' In every community people take for granted certain conventions about eating: even the usual number of meals varies from one to four or more. Sometimes custom reflects useful practices of the past. In North America, turkey (*Meleagris gallopavo*) and the pumpkin (*Cucurbita*) were valuable foods for the native Americans and for the early European settlers. They are still produced and eaten, but largely for their symbolic significance at Thanksgiving.

Other customs reflect a more remote past. At Easter, in many communities, special cakes ('hot cross buns') are baked and eaten. They celebrate the vernal equinox. This, the moment in spring when day and night are equal, was an important event for primitive cultivators. The cakes are believed to be derived from ancient Egyptian and Phoenician offerings to Astarte, goddess of fertility. The Saxons of Western Europe, early in the present era, similarly marked their loaves with crosses in honour of Eastre, goddess of dawn.

Turkeys, pumpkins, loaves and buns are of course nutritious. The same cannot be said of the solemn ritual, practised by some Christians, of consuming a wafer and a sip of wine on Sunday mornings: these are no substitute for breakfast. Also without homeostatic function are many strict taboos concerning eating. Orthodox Jews must have their meat killed in a certain way; they also require separate sets of dishes for some foods. In India, Brahmins allow only certain classes of people into their kitchens. The English have or had a horror of eating horse meat, although it is nutritious. Sami (Lapps), in the north of Norway, have a similar attitude to reindeer (*Rangifer tarandus*), although for many Norwegians reindeer steak is a delicacy.

Peculiarities of custom can be of great practical importance. A biochemist may wish to tell people how to choose a good diet for themselves and their children. However excellent her advice, it will be ineffective if ritual, custom and prejudice are not allowed for. Such difficulties remind us of the obstacles we meet if we look at ourselves solely as physical systems, such as robots.

Growing Up

The previous pages emphasise the maintenance of steady bodily states, especially in adults. Development from egg to adult displays another kind of stability. Many parents know this. The rapid growth of a healthy infant may be interrupted by a brief illness. On recovery, growth accelerates, back to the weight the child would have achieved if not ill: it achieves a target value. This is an example of developmental regulation (homeorhesis).

The whole of normal development is a progress through an astonishing series of strictly regulated changes. I wish every reader could watch the early stages of a sea urchin. In spring, a female pours about ten million eggs into the sea, a male perhaps a billion sperms. In the laboratory, both can be induced to release their gametes into a dish. Sperms soon surround each egg. One enters and a membrane forms round the egg; no other sperm can now get in. After 30 minutes, the haploid egg and

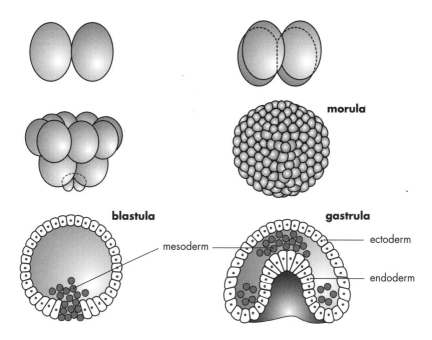

Embryos transformed. In the development of a sea urchin, the first distinct structure is a ball of cells which soon hollows out. Next, a strange reorganisation can be observed by applying harmless dyes to the living embryo: some cells migrate from the outside to the inside to form a bag with two layers, the gastrula. The outer layer (ectoderm) eventually gives rise to the outer part of the skin, the nervous system and much else; the inner, endoderm, forms the lining of the gut and the lungs, among other things. Another migration forms a third layer, the mesoderm, between the first two. The mesodermal cells put out fine processes (filopodia, not shown), with sticky tips, and crawl over the other cells. They are the source of the muscles and skeleton. The two bottom figures are shown cut in half.

sperm nuclei have fused and the respiratory rate rises to six times its previous level: extra energy is released.

Within an hour, the new nucleus begins to divide and soon the whole egg divides into two; next the nuclei and cells divide again. If the four cells are now separated, each can develop into a complete animal. They are totipotent or capable of anything. Or, if two two-celled embryos are put together, they can develop as a single large embryo. Similar early flexibility is found among some mammals. The Texan nine-banded armadillo (*Tatusia novemcincta*) regularly gives birth to uniovular quadruplets. Some multiple human births too originate from a single fertilised egg: 'identical' twins are uniovular.

All chordates, including ourselves, have early stages which, to some extent, resemble those of a sea urchin. All form the three layers, outer, middle and inner, shown above. Many features, however, depend on how the embryo is fed. Ova are rather large cells with more protein than usual. Those of mammals may be one-tenth of a millimetre in diameter, nearly ten times the dimensions of a liver cell. The largest, those of ostriches (*Struthio camelus*) and some sharks, are gigantic, with a diameter of about 80 millimetres; this is nearly all food reserve. The last figure refers only to the 'yolk': what, at breakfast, we call the egg of a domestic fowl includes outer layers—the white and the shell—which are not part of the egg cell itself. The white, or albumen, is mainly a store of water. The porous shell protects the growing embryo.

The whole of an enormous ovum cannot divide. The development of a bird is therefore distorted. The nucleus is in a patch of cytoplasm on the surface of the yolk. Fertilisation is internal, before the yolk is covered by white and shell. After fertilisation, the nucleus and cytoplasm divide repeatedly to make a disc of cells in which an embryo forms. Cells at the edge of the disc move outwards and cover increasing areas of the yolk. About 30 hours from the beginning of incubation, a neural tube (precursor of the central nervous system), a notochord (precursor of the vertebral column) and muscle blocks are visible. The front of the neural tube is expanding to form a brain; a heart, blood vessels and blood have developed and are functioning. Cells on the yolk surface digest the yolk; the products of digestion are carried in the blood and supply the embryo.

Mammals evolved from reptiles with large eggs, like those of birds, but their embryos are differently nourished. A mammalian egg cell moves from the ovary down the fallopian tube. If the female has been inseminated, the egg is entered by a sperm and begins to divide while still on the move: it already has a distinct structure when it becomes implanted in the wall of the uterus. There a placenta develops, partly of embryonic cells, partly of uterine tissue. The two blood systems are not continuous: substances pass across thin membranes which separate the blood of the mother from that of the embryo. The embryo itself develops rather as if it had the massive yolk of its ancestors: it forms a disc of cells in which the main structures appear in much the same way as those of reptiles and birds. It

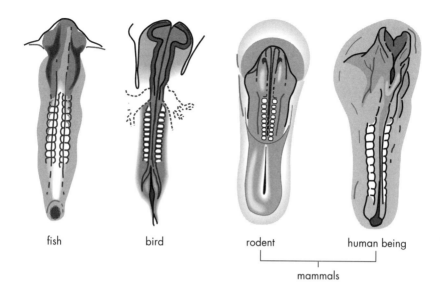

Early embryos seen from above. The first two are lying on large masses of yolk. Mammals have only a small, vestigial yolk sac but develop in a similar way. In each case, the first divisions of the fertilised ovum give rise to a sheet of cells. In this, by migration, arise the beginnings of the nervous system, blood system, muscles and other structures. The pairs of muscle blocks in each embryo can be seen; in the centre is the neural tube (the beginning of the nervous system), with the beginning of the brain in front.

even has a small yolk sac but without yolk. Here is a notable example of recapitulation of an ancestral feature (chapter 2).

A small mammal develops in the uterus for three weeks or less. At the other extreme the gestation period of an African elephant (*Loxodonta africana*) is nearly two years. Human pregnancy lasts about 280 days (or nine months) but is inconveniently variable.

MILK

All infant mammals depend at first on the secretions of mammaries which are modified sweat glands. Mammals vary, however, both in the composition of their milk and in how often the females deliver it. Tree shrews (*Tupaia*) of South East Asia visit their young every two days; like whales, they secrete a very concentrated milk. Similarly, rabbits (*Oryctolagus cuniculus*) allow

Human development: timetable before birth

Week from last menstruation

2	Fertilisation
3	Egg moves down tube into uterus; begins to divide
4	Implantation in wall (endometrium) of uterus
5	Early stages of skeleton and nervous system
6	Head, heart, tail appear; 'gills' develop; rudiments of arms and legs; 6 mm
7	Chest and abdomen; eyes, fingers and toes developing; 12 mm
8	Face and external ears developing; gill rudiments disappearing; 21 mm; 1 g
9	Face fully developed; resembles human child; now called a fetus; 30 mm; 2 g
14	Limbs, fingers, toes, nails well formed; external genital organs; 77 mm; 30 g
18	Movements begin; heart can be heard; hair all over body; 190 mm; 180 g
23	Head hair; 300 mm; 450 g
27	Eyes open; 350 mm; 875 g
32	If born can survive with special care; 400 mm; 1425 g
40	Full term; much head hair; 500 mm; 3250 g

All measurements and weights vary greatly.

their young to suckle only once a day: each gulps the milk quickly and swells up like a little balloon. At the other extreme, the milk of Primates is very dilute: suckling monkeys and apes are carried all the time and are fed on demand. Human milk, like that of other Primates, is dilute.

A woman's breasts begin to enlarge during adolescence, owing to growth of fatty tissues. Their size is unrelated to the capacity to produce milk. During pregnancy they enlarge further. A baby can suck at the breast within a few minutes of being born; at this stage the breasts secrete a mixture (colostrum) with a high proportion of protein, much of which consists of antibodies (immunoglobulins), and an array of other antibacterial substances; also present are many white blood cells of several kinds. All these combine to protect the infant from infection. After the first day or two, normal milk begins to be produced. (Sometimes it is produced in excess. It can then be frozen, stored in a hospital and given to premature babies.)

> **BREAST FEEDING**
>
> One of our strangest customs, common in rich countries, is to feed infants on imperfect, even dangerous, mixtures from a bottle instead of the marvellous secretion produced by almost every mother. Here therefore are some advantages of breast feeding.
>
> - The composition of human milk is exactly right for human babies. No 'formula' equals it. Moreover, its composition alters during a feed, to match the infant's need: it is concentrated at first, later dilute.
> - Breast milk, like colostrum, contains antibodies and white cells which protect the infant from disease, including allergic disorders. It also stimulates the development of the infant's own immune system.
> - Breast fed people are less likely than those who are bottle fed to become very fat and so are protected from the illnesses which go with obesity.
> - Breast feeding is much less trouble and less expensive than bottle feeding. Unlike milk from a bottle, it is not a source of infection.
> - Breast feeding strengthens the emotional bond between mother and infant; infants fed on demand and kept warm cry rather little.
>
> *It follows that lactating women should be treated as privileged persons in the community.*

THE CONSTANCY OF NUCLEI: CLONES

The cells of an early embryo all look and are much the same. Each has, in its nucleus, a complete set of genes. From this beginning, a human being acquires about 250 distinct kinds of cell. Other species have similar numbers of cell types. What happens to their nuclei? Studies of the rather large eggs of amphibians, such as a frog (*Xenopus laevis*), have told us much about the cell nucleus, with all its genes, in development. In the experiment shown overleaf, a fertilised ovum has its nucleus destroyed by radiation. Another nucleus is then taken, by micromanipulation, from a *fully differentiated* skin or blood cell and inserted in the egg. The insertion sets development going;

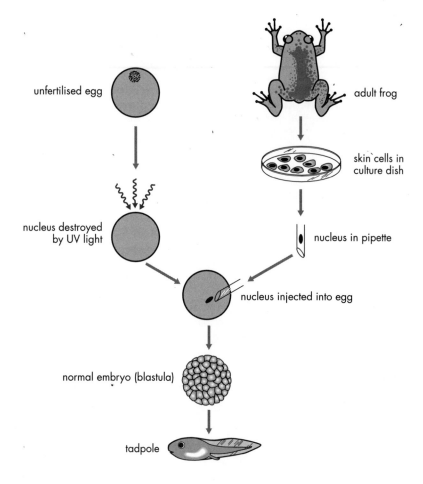

A nucleus from the skin cell of a frog is transplanted into an ovum from which the original nucleus has been removed. The ovum develops into a normal tadpole. In this way the nucleus of a fully differentiated cell is shown to retain the whole array of the organism's genes.

and the egg, with its substitute nucleus, develops into a normal tadpole. The new nucleus, *from a specialised cell*, proves to be just as effective as the old one.

Clones are a consequence of this consistency. 'Clone' is derived from a Greek word meaning 'twig'. It referred originally to plants propagated by cuttings, hence without sex; but it now has additional meanings. Some animals can multiply asexually, including some Coelenterata (the jellyfish phylum) and the flatworms. The basic structure of coelenterates is of a two-layered bag into which prey is dragged by tentacles surrounding the mouth. *Hydra* is a rather weird example which can reproduce by budding. If it is cut in two or three, each part reforms into a whole individual. The two or three make a clone. The presence of a full complement of genes in each cell makes this reorganisation possible.

The flatworms have three cellular layers instead of two. Most free-living forms (planarians), are only a few millimetres long. They can reproduce asexually, by dividing transversely into two. An experimenter who has only one planarian but wishes for several may cut the one into several pieces. Each piece then becomes a complete worm. The products again form a clone.

Today, beginning with one cell, a colony of cells can be kept and grown in a nourishing fluid (tissue culture) to produce a cell clone. The reader, as a child, may have put the top of a carrot in water and watched the growth of leaves. The carrot can be used in more elaborate experiments. A portion is cut out and the cells are separated. One cell is then grown to form a clump of multiplying cells which become an embryo. From that cell clone, roots and leaves arise and, eventually, another carrot.

It is now possible to propagate mammals from single cells. The latest news is of cloning sheep and monkeys. Cloned sheep can be of economic use. If they are all reared in the same conditions, they are likely to be very similar. They are not, however, all the same. The possibility of cloning human beings is also being debated. Some people have the notion that a cloned person would be identical with the individual from whom the cell nucleus was taken; but in fact a cloned human being would be a distinct individual, not a replica of another person. The idea that a man or woman can be replicated by cloning is an example of the tendency to attach excessive importance to genes: it ignores the environment in which a person develops. More important, many people are so repelled by the idea, that proposals to perform such experiments are meeting massive opposition.

How Do Cells Become Different?

If all cells are equipped with a full complement of genes, how do they become different? This has been, and remains, one of the most formidable of biological problems. Delicate experiments have, however, shown that the position of a cell in relation to others can decide its fate. This process, induction, is revealed by shifting groups of cells around. Ectodermal cells which would ordinarily become skin can be moved into a position, inside an embryo, where they develop into quite different cells, such as those of muscle. Another such experiment concerns the development of the eye. The seeing part of the eye, the retina, originates as an outgrowth of the brain, the eye cup. It is covered by ectoderm, most of which gives rise to skin. But, over the eye cup, the ectodermal cells form a lens. Any ectodermal cells can be moved into this position and become a lens: the crucial requirement is contact with an eye cup.

Very rarely, an ovum or sperm begins to develop while still in its gonad. It then gives rise to a teratoma—a disorganised mixture of skin, bone, gland and other tissues: the gonad does not provide the regulatory effects needed for normal development. Yet the individual cells of a teratoma are normal. Cells can be taken out of a mouse teratoma and implanted in an early embryo. There they join with the cells already present and develop into part of a normal mouse. Again differentiation is shown to depend on the conditions in which the cells develop: in this case the transplanted cells are acted on by the embryonic cells already present.

Extremes of precision are seen in the development of the nervous system. Some nerve cells begin life in a structure, the neural crest, on each side of the neural tube (shown on page 113) but move along fixed paths to distant destinations. Some become pigment cells in the skin. Others form part of the adrenal gland or nerve ganglia in the abdomen. Yet other nerve cells put out long axons which too find their way to remote targets. The cells of the target organs release nerve growth factor and other substances which attract the growing processes. These substances are also important in the adult. When tissues such as skin or muscle are injured, the fine processes of nerve cells are cut; their regrowth is then stimulated by nerve growth factor.

The growth of some nerve cells depends also on external stimulation. The connections between the retina and the brain (shown on page 128) develop normally only if the cells in the retina are stimulated. Such findings explain one kind of blindness. Some young children cannot see with one eye owing to cataract: light to that eye is cut off. If the cataract is not treated quickly, the result is permanently defective vision on that side: without stimulation, the cells in the optic nerve have failed to make the necessary contacts in the brain. For similar reasons, a squint must be corrected early. Without practise at using two eyes together, the child does not develop stereoscopic vision.

According to a rather astonishing recent report, external stimulation early in life does more than stimulate the growth of nerve processes: it even provokes the multiplication of nerve cells. We have here yet another example of the importance of interaction with the environment during development.

Abnormal Growth: Tumours

Sometimes, as in a teratoma, tissues grow when they should not and produce a tumour (or neoplasm). The simplest merely enlarge and are called benign. Familiar examples are warts, which grow from skin cells. More serious is the excessive growth of the prostate gland, common among elderly men, which may obstruct the release of urine and require surgery or other treatment. Benign tumours of the ovary, which can grow to a vast size, also have to be removed.

The most serious tumours are the malignant growths or cancers. They invade other tissues, stimulate the growth of blood vessels and so acquire a good blood supply; some, as a result, grow rapidly. Many also release cells into the lymph or blood and set up colonies (metastases) elsewhere in the body: a cancer in the kidney can generate secondary growths in the bones and lungs. Hence removing the primary growth may not cure the condition.

Most cancers, unlike teratomata, arise from differentiated tissues, especially those that continue to grow throughout life. They include the malignant melanoma of the skin common among whiteskins after prolonged exposure to the sun. Another is primary carcinoma of the lung, most of which is due to smoking cigarettes. Yet others, the leukemias, can result from exposure to ionising radiation such as X-rays: stem cells in the bone marrow, which normally give rise to blood cells, no longer do so but instead may produce more stem cells; the cells in the

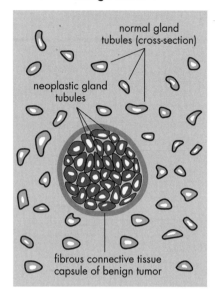 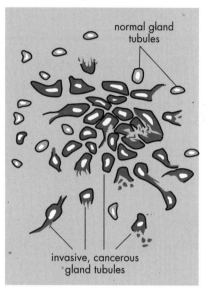

Tumours. On the left, a benign growth, derived from gland cells, enclosed in a capsule. On the right, a cancer also derived from gland cells (an adenocarcinoma): the cancer cells invade the normal tissues.

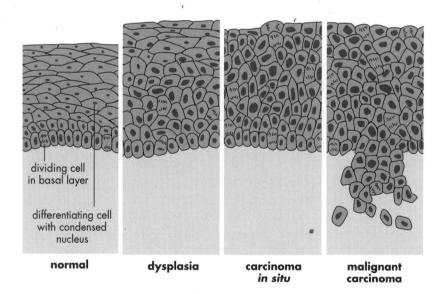

From left to right, stages in the development of a cancer of the human uterine cervix. On the right, malignant cells are invading the connective tissue below the epithelium. Distinguishing these stages is important for diagnosis and prognosis.

blood, especially the red cells (which have short lives), are then not adequately replaced.

Each abnormal growth probably arises from a single cell. Usually, a gene in one cell must first mutate. But this is only a first step: additional mutations are needed; in addition, later stages of some cancers require infection with a virus; others require chemical stimulation. Usually, for cells to become malignant, they must have accumulated the effects of about six such rare events. Sometimes distinct stages in the development of a cancer are recognisable under the microscope: cancer of the uterine cervix, illustrated opposite, is an example.

Hence a long time usually elapses between exposure to a carcinogen and the appearance of a malignant growth. When, in 1945, atomic bombs were dropped on two Japanese cities, they released ionising radiation. After five years, leukemia greatly increased among the survivors; the peak was reached after eight years. Heavy smoking causes lung cancer only after ten or more years. When workers in industry are exposed to carcinogens, the delay may be still longer: the mesothelioma due to asbestos takes twenty to thirty years. It is easy to see why most cancers occur in old people.

Probably, more than 80 per cent of human cancers are due to identifiable features in the environment. They are therefore preventible. Some cancers clearly need environmental agents for their development even when the agents have not been identified. Cancer of the stomach is common in Japan but rare in some African countries, such as Uganda. Also, Japanese women in Japan have a much lower incidence of breast cancer than white women in the United States. Such variation could be due to genetic differences between populations; but the cancers among emigrants resemble those of the people in their new country: the differences must therefore be environmentally caused.

Despite the new knowledge of tumour cells, the most important kinds of malignancy are still increasing: not only those of the lung but also of the breast, the prostate and the intestine. Because of the long series of events between the initial action of a gene and the development of disease, discoveries of 'genes for' various kinds of tumour have rarely been helpful. Moreover, the cells of an abnormal growth vary and often mutate; each kind of growth therefore consists of a mixture of cells, some of which multiply more rapidly than others. Worse, a growth, such as a lung carcinoma or a melanoma, in one person may differ biochemically from the same kind of growth in another.

Today, therefore, treatment of cancers continues to be surgery, chemotherapy and related methods. For the longer term, however, the most important measures are in the realms of human ecology and public health. The external agents which provoke malignancy can be removed or avoided. In some industries, this has already been done. Cancers due to asbestos in insulation, benzene in paint, mineral oils in metal work and many others, formerly common among exposed workers, are now rare. But prevention is not only a matter of avoiding carcinogens. A new finding concerns nutrition. Eating plenty of fruit and vegetables goes with a low incidence of cancer.

That last sentence summarises the results of research on a large scale, spread over many years. Like other findings cited above, it belongs to epidemiology. This is not a glamorous subject and, unlike genetics, it receives little publicity. As we know also from chapter 4, it has, however, the merit of providing knowledge which is widely useful now.

Epigenesis and Adaptability

Emphasis on lifestyle brings us back to the interaction of genotype and environment. The orderly and predictable development of complex organisms occurs only in certain conditions. If a fertilised egg is put in unusual surroundings and survives, its development may be grossly abnormal. Human beings have well developed limbs, each with five fingers or toes. Yet, when a drug, thalidomide, was given to a number of women early in pregnancy, these apparently fixed traits did not develop: many children were born with only short stumps instead of arms or legs. The drug changed the conditions in which the embryos were developing. This calamity represents an extreme case of an environmental influence on apparently universal features. Other structural effects are less drastic and more familiar; some are welcome. The rise in average stature, during the twentieth century, among the poor in many countries, was due to improvement in the environments in which children were reared. Such variation in development is a general phenomenon among both animals and plants.

One kind of individual variation, which we see every day, is a result of adapting behaviour to the demands of the environment. The ability to respond flexibly to 'time and chance' may seem to conflict with the findings on fixity of development in the nervous system. The brain which grows with such marvellous precision is the source of behaviour and, as we see in chapter 9, much animal behaviour is stable in development: it is 'species-typical' and highly predictable. But the brain, once developed, is not only a generator of such stereotyped activities: especially in the most complex animals, and above all in human beings, it makes possible the individual adjustments to circumstances which we call intelligent.

CHAPTER 7

BRAINS AND MIND

> Whence then cometh wisdom?
> And what is the place of understanding?
>
> THE BOOK OF JOB

THE HUMAN BRAIN IS the site of intelligence; it enables us to learn (and to forget); it is the source of consciousness. For centuries, physiologists and philosophers have been baffled by its complexities. Teachers and others have therefore resorted to extreme simplification. Uncounted innocent school children have been encouraged to think of the brain and the nerves as an electrical system: nerves have been equated with wires and nerve cells, perhaps, with switches. In 1976, the British Broadcasting Corporation put out an authoritative series of lectures, by Colin Blakemore, entitled *Mechanics of the Mind*. The cover of the printed version shows a human head full of wires and equipment. The many attempts to explain intelligence and learning by undemanding theories have influenced popular attitudes and even the policies of governments. Meanwhile, much has been learned about the real brain and about human abilities.

NERVES AND BRAINWORK

We begin with the easiest part: some facts about nerves. As a preliminary, I suggest that the reader tries pricking a toe or

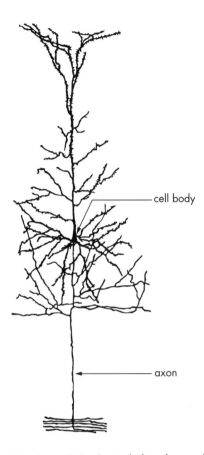

A vertebrate nerve cell. The cell body includes the nucleus. Many short processes (dendrites) make synaptic connections with hundreds of other cells and carry incoming impulses. The long axon carries outgoing impulses. Not all nerve cells have long axons.

finger with a pin (or asking a friend to do it). You should concentrate very closely on the resulting sensation: it is in fact double. The double pain response (which is only very slightly painful) is caused by the transmission of impulses to the brain, from sense organs in the skin, by two kinds of nerve fibre which carry impulses at different speeds.

Nerve cells (neurons) have the same main components as other cells, but they are usually elongated and have a branching structure. The processes (nerve fibres) of neurons conduct signals. An individual nerve impulse (action potential) consists of a sequence of chemical changes accompanied by a change of electrical potential. The number of impulses passing along a fibre in a second may be many or few, according to the intensity of the stimulus. The smallest fibres (axons) of a mammal conduct at about one metre a second, the largest, at about 120. The process is therefore quite unlike electrical conduction. To write of brain 'circuits', or of neural systems as 'wired in', gives a quite wrong impression.

Nor is the brain anything like an electrical machine. Exposed by a surgeon, a human brain is a gelatinous mass of which the parts are easily displaced. About two-thirds by weight is salty water—an excellent conductor of electricity, hence an electrical engineer's nightmare. It consists of nerve cells and other components. The processes of the neurons end close to the surfaces of other neurons and so form synapses. One neuron acts on another by secreting substances which cross the gap and cause the second neuron either to fire (excitation) or to stop firing (inhibition). A mammalian brain and spinal cord consists largely of interneurons which are in synaptic relationship with many hundreds of other neurons.

The relationships of the parts of the nervous system are illustrated in the diagram of the knee jerk, familiar from its use in diagnosis. A tendon reflex is an apparently simple example of nerve action but, as the picture shows, it is not in fact simple. Still less simple is what you will shortly do: you will get up, put this book away and move to a new activity. In doing so, without thinking, you will employ, with elegant exactness, dozens of muscles and millions of nerve cells. The exactness depends on the cerebellum. At all times, signals are reaching the brain and spinal cord from the limbs and other mobile parts. The cerebellum processes this information: its output regulates balance and movement. A mammal with a damaged cerebellum may be unable to run without falling over. A test of human cerebellar function is to ask a person to close the eyes, to extend an arm and to touch the tip of the nose with a forefinger. A normal person can do this promptly and accurately. From an engineering point of view, considering all the structures involved, this is quite an achievement; yet it is nothing to the speed and precision shown by, say, a musician or even by the person typing these words.

How the cerebellum works is not known. The same applies still more to the cerebrum or cerebral hemispheres. The outer part, or cortex, of the hemispheres consists of many millions of

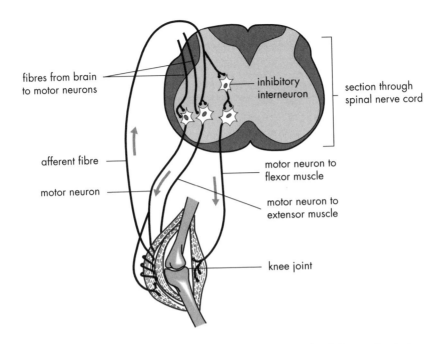

The structures involved in a reflex such as the knee jerk. If the tendon below the kneecap is hit, the muscles in front of the thigh contract and produce a brief kick. Signals have gone to the spinal cord from stretch receptors in the tendon. Nerve cells in the cord are stimulated and evoke impulses in motor neurons whose axons go to muscles in the front of the thigh. Meanwhile, the muscles in the back of the thigh (the antagonists of those in front) relax; this requires inhibition of certain neurons in the spinal cord. The process depends on a two-way system of connections among spinal cord interneurons, usually (and mercifully) left out of textbook diagrams.

nerve cells arranged in layers. Some regions, the primary projection areas, receive impulses from particular sense organs. The optic nerves, shown overleaf, run from the retinae of the eyes to the thalamus which lies underneath the cerebrum; and tracts of nerve fibres run from the thalamus to the back of each hemisphere. The retinae are represented, point by point, in this region (the occipital or visual cortex). A small injury in the visual cortex of a human being results in a corresponding small blind spot in both eyes. From there, however, connections are made to several other parts of the cerebral hemispheres: point-for-point correspondence then breaks down.

Another region of the cortex, the somatic sensory area, receives a massive input from the skin (including the impulses of the double pain response). Next to it is the motor cortex. Here are many large cells (motor neurons), each with a long

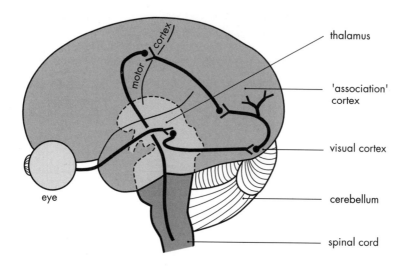

Connections in the brain of a mammal, between the eye and the output to the muscles. Impulses travel from the retina to the thalamus; the thalamus relays them to the visual cortex; from there connections go to several other parts of the cortex and eventually to the motor region. The last contains large cells with axons running to the spinal cord; there they make contact with neurons of which the axons go to the muscles.

A diagram like this makes the central nervous system look simple. It is not: at each stage, each neuron makes contact with hundreds of others. In the cortex, many regions are involved. Every stage has neurons which conduct impulses in the direction opposite to that shown: they evidently carry regulating, inhibiting and dampening impulses.

axon running to the spinal cord; there they end on neurons which, in turn, send axons to the muscles.

Intervening between the sensory areas and the motor cortex is a system of barely describable complexity. The human brain is estimated to contain about 100 000 million (10^{11}) nerve cells. The best way to comment on it is to remind the reader of some of the things we do with it. You are probably not a chess grandmaster, a leading solo violinist, a mathematical genius or even an international sports star. Nonetheless, what you do when you use your brain every day is remarkable enough. These achievements are so familiar, it is easy to overlook how remarkable they are.

One is the ability to recognise an object from any angle and over a great range of distances. The pattern falling on the retinae of our eyes, when we see, across the road, a friend or even a simple sign such as BUS STOP, is quite different from that of the same friend or sign when seen nearby or at a different angle. You are now using this ability at great speed while reading. Individual letters are identified, and words understood (I hope), regardless of their size or of the typeface. Similarly, a tune can be identified regardless of the key in which it is played, the instrument used or the tempo of the performance. And, still more important, we understand a sentence spoken in a variety of accents and uttered in falsetto or basso profundo. These are examples of stimulus generalisation. They represent a quite unmachinelike tolerance of variation in the stimulus which provokes a particular response.

Similarly, we find response generalisation: a pianist can play a chord, marked on a sheet of music, by any of several combinations of the ten fingers. The reader can, with an effort, write legibly with the hand not normally used or even with a foot.

Yet an account of such everyday achievements seriously understates what some people can do. The variation we find in human abilities is highly disconcerting, notably that among outstanding persons some of whom have enriched the lives of millions. The achievements in music of W.A. Mozart (1756–1791), in mathematics of Isaac Newton (1642–1727) or in philosophy of Plato (428–347 BC) are quite beyond the scope of even most highly competent persons in these fields. Their brains were no doubt different from ours, but we have no clue to what the differences were. The brain of a giant of theoretical physics, Albert Einstein (1879–1955), has been preserved in formalin. It would be naive to suppose that anything useful can be learned from studying it.

Localisation, Psychosurgery and Criminal Assault

If the brain were a machine, we could usefully ask which parts are responsible for particular functions. The cerebellum, we know, is especially concerned with regulation of movement and balance; and small regions of the hypothalamus seem to play a special part in controlling temperature and appetite. In addition, the primary projection areas of the mammalian cerebral cortex have clearly defined functions.

But what of the rest of the brain? This subject began to be seriously discussed in the eighteenth century. Then the debate collapsed: according to a strange story, widespread in the nineteenth century and later, a person's character could be identified by the 'bumps' on the skull. This doctrine, phrenology, influenced colloquial speech: people referred to their bump of direction for the ability to find their way; or to their bump of reverence for willingness to accept the word of authority.

For a long time, authentic knowledge of localisation came from the effects of small injuries due to accident, rupture of a blood vessel or to violence, as in war. But now new methods allow observation of changes in intact brains: it is possible to record small local fluctuations in temperature and metabolism which go with extra activity.

Much new work has been on vision. When we see an object, we observe its colours, shape and movements as a single experience. This is an aspect of that mysterious state of awareness called consciousness. In the brain, each of these attributes seems to be processed separately. A person with an injury in one small part of the cerebral cortex may be unable to see or to remember colours (achromatopsia), although in other respects vision is unaffected. Worse, in akinetopsia the ability to see objects in motion is lost: objects are seen and understood normally while they are still; as soon as they move, they disappear.

These defects are sensory. Others are 'motor'. Large injuries in one region of the brain cause major deficiencies in the ability to speak. Small lesions can lead to loss of particular aspects of speech: inability to say words correctly may result in errors such as 'loliphant' for 'elephant'; or spoken sentences may be ungrammatical.

Yet other injuries affect memory. The existence of several kinds of memory is familiar. We can recall, say, where a pen was put down a minute ago; we can (usually) identify the place where we parked a car six hours ago. Both are, as a rule, soon forgotten. But, at the other extreme, an old person's memories of childhood may have been retained for more than 70 years. Damage to the temporal lobe of the brain on both sides can cause total loss of longterm memory. A patient in this state can have a sensible conversation with a nurse or doctor but fail to recognise them a day later. On all this evidence, many functions in the human brain are highly localised. But we do not know how the operations of all these regions are combined to unify our understanding and our actions. We have no comprehensive physiology of thinking (or of failing to think). We do not understand consciousness.

Ignorance, however, has not prevented rash and destructive action. In an earlier period, insane persons were often treated by exorcism, during which devils were supposed to be cast out. (Occasionally, they still are.) In the 1940s, a new procedure was introduced: abnormal thinking or mental illness was attacked with surgeons' knives. The kind of psychosurgery most used was frontal lobotomy or leucotomy: part of the front of the cerebrum was destroyed. By a crude analogy, such an operation might be regarded as disconnecting parts of an electrical system. Many physicians were shocked by this procedure, partly because it has no valid foundation in physiology or in clinical research. Yet, especially in the United States, the operation was performed on thousands of persons, rich and poor. The victims included not only seriously disturbed patients but also people described as depressed, anxious or merely 'emotionally ill'; some were children diagnosed as overactive. A common outcome of lobotomy is 'vegetabilisation': the subject becomes passive and unenterprising but is still capable of performing simple tasks such as housework.

The macabre story is told of a patient who was being treated in this way under local anesthetic. As the cut was made, he was asked, 'What is going though your mind?'. He replied, 'A knife'. A proper description of such operations is criminal assault leading to grievous bodily harm. They represent a willingness to reduce the human personality to a mechanism to be tinkered with. They warn us (again) not to regard our fellow beings as robots.

CONDITIONAL REFLEXES AND PRESSING LEVERS

Faced with the enigmas of brain functioning, an outstanding American psychologist, K.S. Lashley (1890–1958), after a lifetime of research, declared that learning was impossible; yet, he conceded, it does occur. During his time and later, strenuous efforts were made to demystify learning or intelligence by experiments on closely confined animals. Much was found out;

but the outcome reminds us of an unkind saying: that every complex problem has a solution which is simple, neat and wrong.

Here is a small observation. Put a docile laboratory rat on a table or bench. It will move around, sniffing and will poke its nose into any accessible crevice. Animals not only exert themselves to find food, shelter, a mate and other necessities: they also explore. What causes this behaviour? What is its function? Major advances have come from trying to answer such questions. But, for the present, I leave the reader in suspense about them because, for decades, the most influential workers in this field ignored them. Far-reaching conclusions were, and still are, drawn from experiments in which the animals either could not move about or, if they could, it was not recorded. And the conclusions have been applied to humanity.

A leading figure was the celebrated Russian, I.P. Pavlov (1849–1936), who began to be trained as a priest but turned to medicine. Pavlov was a brilliant experimenter. By ingenious investigations of dogs he became a founder of the modern knowledge of digestion. In middle age, when already renowned as a physiologist, he turned to behaviour and became still more famous.

Pavlov continued to use dogs and physiological methods. He hoped to show how the brain controls intelligent conduct; yet, during his experiments, the dogs were not taught skills or even house trained. His findings were all on conditional reflexes (CRs). CRs have often been thought of as products of an electrical system, with switches which go on and off according to circumstances. Yet their principal interest is in the differences they reveal of animals and people from machines.

Pavlov's dogs were first trained to stand on a table in a harness and to accept a tube for collecting saliva. When a trained and hungry dog is on the table it hears a sound. It pricks its ears and turns its head and the heart and respiration rates rise (the orienting reflex). So the dog shows signs of interest, agitation or emotional arousal. These signs are usually left out of descriptions of CRs.

Food is presented, the animal eats and the amount of salivation is recorded. With repetition of this sequence, the orienting reflex disappears and the heart and respiration rates no longer rise. Instead, the dog, on hearing the sound, looks to where the food will appear, paws the ground and licks its lips. It is expecting nourishment. Correspondingly, it begins to salivate *in advance* of the food. Eventually, salivation is induced by the sound alone. Hence the salivation after training is something new: it occurs on its own, before the food.

If now, on a series of occasions, the dog hears the bell but receives no food, the learned salivation fades and disappears (extinction). But suppose that extinction has been made to occur on one day. The next day, the dog is presented with the same conditions as before. It then salivates. The response acquired during the original training is largely restored. Extinction is not forgetting: it is a learning *not* to respond which fades quite rapidly.

During years of experiments, Pavlov and his successors revealed many other disconcerting, unmachinelike complexities. One is the effect of a confusing situation. A dog is trained to salivate when presented with meat on one dish; it is also separately trained to salivate to biscuits on a quite different dish. It is then offered meat on the biscuit dish and biscuit on the meat dish. The salivation fails; and the bewildered dog looks from one dish to the other. Once again, we see a result of what the animal expects.

The most famous complication of CRs is the 'experimental neurosis'. A dog, strapped down as usual, is first trained to salivate to a tone. It is also trained *not* to salivate to a tone of a much higher pitch. In repeated trials, the second sound is progressively lowered until it is difficult to distinguish from the first. The training then fails and the dog may resist being taken to the laboratory; or it may go off its food altogether and sit miserably in a corner.

In short, the behaviour studied by Pavlov illustrates complex effects of hunger and emotional arousal in an animal prevented from moving about: it is nothing like that of a machine; nor does it represent the behaviour of an animal (still less, a human being) moving freely. And, despite what Pavlov hoped, it does not tell us how the brain works.

Yet the idea of the conditional reflex had, and still has, a tremendous impact. The study of CRs was long supposed to solve major problems of learning and intelligence. Fiction gave Pavlov yet another dimension: garbled versions of Pavlovian conditioning appear in stories in which villains attempt psychological control over their victims. Meanwhile, critics ridiculed the whole enterprise. Here is a passage from an irreverent fable, *The Adventures of the Black Girl in Her Search for God*, by George Bernard Shaw (1856–1950). The heroine meets Pavlov, who says,

> 'If I give you a clip on the knee you will wag your ankle.'
> 'I will also give you a clip with my knobkerry; so dont do it' said the black girl.
> 'For scientific purposes it is necessary to inhibit such secondary and apparently irrelevant reflexes by tying the subject down' said the professor. 'Yet they also are quite relevant as examples of reflexes produced by association of ideas. I have spent twenty-five years studying their effects.'
> 'Effects on what?' said the black girl.
> 'On a dog's saliva' said the myop.
> 'Are you any the wiser?'

As the black girl implies, the story of conditional reflexes contains an absurdity: to understand human behaviour, we should concentrate not on the dog but on the experimenter; his behaviour, however, is even more complicated than that of a freely moving dog.

The other leading attempt to simplify learning was that of the American behaviorist, B.F. Skinner (1904–1990). In his system, all our learned behaviour can be explained by the immediate consequences—pleasurable or painful—of what we do. His subject (usually a monkey, a rat or a pigeon) is put in a container (a 'Skinner box') where pressing a lever or pecking a target leads to delivery of food or water into a small bowl; or it may switch on a light or heat, or switch off a source of discomfort. If the animal is hungry or thirsty, it will learn how to get food or water. Similarly, an animal can learn how to avoid injurious stimuli.

Only depression of the lever is recorded. For Skinner, such cramped procedures represent the only scientific psychology. He and his many followers believed that his doctrine can solve problems of industrial production, mental health, teaching and politics. Managers have studied it in the hope of increasing profits; psychotherapists have used it for treating mental illness; and teachers have furnished their classrooms with machines inspired by Skinner's system. Skinner himself has outlined an ideal community managed by a ruler who administers rewards and punishments as required.

Skinner's ideas, however, are plausible only if one disregards facts. In industry, people are not influenced only by the rewards of pay, soft music and coloured walls: production is improved if workers have some say in what they do, that is, if they are treated as people, not as circus animals or as pairs of hands. Pupils in schools, too, notoriously respond as individuals to the teaching they receive. As for mental illness, some therapists have tried to use the behaviorism of Pavlov or Skinner. They have then found that patients have thoughts, feelings and wills of their own: like workers and school children, patients behave as persons, not as systems with restricted software; recovery is influenced by what they are told and by the attitudes of nurses and physicians, as well as the treatment itself.

Why has behaviorism nonetheless been widely accepted? Management, teaching and psychotherapy are full of uncertainties. Conditioning and Skinner's 'behavioral engineering' at first gave their practitioners confidence of the kind felt when one uses efficient machinery. It removed doubt and anxiety. Behaviorism also fits our age of manipulation. Today, a person who has a new idea or product or policy tries to sell it by advertising. This is quite different from persuading people by argument. Manipulating people has become a major profession which gives great help to those who want profits or power.

From which you may suppose Skinner to have been an ogre who regarded us as puppets, mechanically operated by the immediate causes of our actions. The reader may indeed be horrified to learn that Skinner used to experiment on his daughter, Deborah, when she was very young. So let me sum up by quoting one of his anecdotes. He was rubbing Deborah's back when she was four years old; and, as he puts it, he decided to test rubbing as a reinforcer (or reward). He waited until she raised her foot slightly and then rubbed. She lifted her foot again and again he rubbed. Skinner writes:

> Then she laughed. 'What are you laughing at?' I said. 'Every time I raise my foot you rub my back.' She had quite precisely described the contingencies I had contrived.

But Skinner does not analyse his daughter's laughter, or her comment or his, in terms of contingencies of reward: only the lifting of her foot. His daughter, like Skinner himself, emerges as a human being.

Working for Change

The simplicities of the behaviorism of Pavlov and Skinner disregard a central feature of animal as well as human behaviour. Much activity achieves not a primary reward, such as food or other comfort, but change or variety. Even in a Skinner box, if pressing a bar switches on a light for a few seconds, the animal learns to do this; if the light is left on, it works to switch it off. It will also work to produce sounds. Such behaviour reminds us of the exploratory activity of the rat described earlier. In new surroundings, mammals (and many other animals) move around as if curious. Monkeys show a liking for novel stimulation also in other ways. Experimenters have allowed them to watch moving objects, such as a toy train going round in a circle; for this, they had to hold a window open and they spent long periods doing so. Chimpanzees also play with and solve simple manipulative puzzles without being rewarded.

Working for stimulation is not impelled by hunger or by any other immediate need. It does, however, contribute to survival. By moving around, an animal learns about its living space. Later, it can use this knowledge when it needs to move promptly from one place to another. As a result, many animals can quickly make new, adaptive responses; previous experience enables them to improvise a short route, not used before, to a goal.

Exploring does still more, especially for the young: it helps them to solve problems in general. Recall the effects of stimulation on the growth of nerve cells in the brain, described in the previous chapter. The innovative Canadian psychologist, Donald Hebb (1904–85), took two young laboratory rats home for his daughters to play with. Later, he tested the rats' ability to learn to find their way through mazes. He also tested similar rats which had been left in the usual boring cages. The animals which had been given more variety of experience did much better. Of course, this was hardly a model experiment. But Hebb's colleagues and pupils confirmed his findings during meticulous studies in the laboratory.

They also discovered effects of different kinds of early experience. Rats have been brought up in cages with walls decorated in black and white. As adults they were trained to go through a door marked with, say, stripes, and to avoid one marked with a circle. They did better than controls kept in featureless cages. Other experiments have been on the manipulative abilities of apes. Young chimpanzees were allowed to play with sticks. They later solved problems which required the use of sticks more readily than did other apes which had not had sticks to play with.

The need and liking for variety and exploration is a feature of everyday human experience: boredom, as Kipling knew, is one of the most distressing of familiar misfortunes.

> The camel's hump is an ugly lump
> Which well you may see at the Zoo;
> But uglier yet is the Hump we get
> From having too little to do.

Its effects have been shown, in an extreme form, in experiments. University students have been paid to lie for long periods on a bed, in a warm, silent room, with hands and eyes covered. Few could tolerate such sensory deprivation even for a few hours. Some developed hallucinations. If, however, exercise was allowed, these ill effects were much reduced. A reader condemned to solitary confinement should take as much exercise as practicable.

Other (harmless) studies have been on young children. It is usual to put a moving object above an infant's cot or pram. The infants enjoy the movements and, as they watch, they are learning to use their eyes and are beginning to interpret what they see. Experimenters have arranged that babies of a few months could *cause* movement in a mobile above them, by wriggling about. They soon learned to do this. At one year, given a familiar toy or a new one to play with, children usually prefer the new one. If they are offered a choice between a familiar place and a strange one, they quickly explore where they have never been before.

Variety of experience helps children to develop their abilities. Psychologists have tested this by imitating experiments first performed with apes. They gave young children sticks to play with; other children saw an adult playing with sticks; yet others played with other things. All were then given the task of using sticks to get hold of an object. The children who had played with sticks solved the problem most quickly.

The more enjoyable stimulation a young child has, the better. Play and curiosity depend greatly on general health and this has a connection with the recent findings on badly fed children (page 106). Such children are less energetic and exploratory than the well fed. They interact less with their surroundings and are slower to develop the skills which depend on exploration.

THINKING: MISS MARPLE AND COGNITIVE PSYCHOLOGY

Exploration and play, as sources of intelligent conduct, take us right outside simple, behaviorist presumptions. The people who try to analyse real human intelligence are the cognitive psychologists. To give a notion of what they have to cope with, I have invented a name, the Marple Principle. The allusion is to Agatha Christie's Jane Marple, an elderly person whose outstanding skill as a detective is owed to experience of life in an English village. Her crucial qualities are a superb longterm memory, and skill at putting diverse items together. In *4.50 from Paddington*, she says:

> One sees a good deal of human nature living in a village all the year round. But my own process of reasoning isn't really original. It's all in Mark Twain. The boy who found a horse. He just imagined where he would go if he were a horse and he went there and there was the horse.

Another name for the Marple Principle is common sense. Unfortunately, intuition and common sense are difficult to understand. It would be helpful if they could be reduced to something comprehensible. Hence the brain has been supposed to be a kind of computer. If so, it should be possible to build machines that think. And indeed, brilliant work has led to the construction of computers which play chess and perform other notable feats. But these machines can operate only within a system of rules chosen by their makers. They are exclusively logical and are therefore far behind the reader and Miss Marple in many abilities.

Here are some differences between you and a computer.

A few pages back, I mention our ability to distinguish objects at different angles and distances. If a machine is to recognise, say, a sphere or a cube, it may be programmed with formulae which specify them. But if we have to identify a shape we do not resort to geometry; and, when we instantly recognise a friend or a bird or a Jaguar XJS, we make use, in ways not understood, of our many memories of such objects.

Consider next the question, asked by the American psychologist, Jeremy Campbell, 'Can crocodiles run steeplechases?' A question of this sort baffles a machine; but you, gentle reader, have no difficulty: because again, without effort, you call up, from your vast store of recollections, mental pictures of crocodiles and steeplechases.

A fundamental difference of a person from a machine is the ability to allow for the context of an event. As I write these words I draw from my own memory something I wrote years ago. The passage begins:

> ... the right elbow is partly flexed, the fingers are extended and the arm is rotated laterally so that the hand moves across in front of the face.

That sentence presents an action as it might be described by a behaviorist or a machine. But I continue:

> This could describe how a gracious personage acknowledges a cheering crowd; but in Australia it is more likely to refer to somebody brushing away a bush fly. We understand both situations when we see them: we can *explain* them.

Our explanations require familiarity with the diverse contexts in which the behaviour appears.

In another difference, we continually transform what we observe: we do not merely record what we see, like a camera; we fit it into a whole set of assumptions and recollections. Although, much of the time, this is necessary, it can also lead to error. Most people, at first sight, read the phrases in the triangles overleaf incorrectly: they *interpret* them. (Look again!)

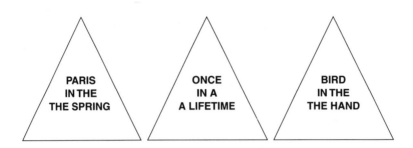

Context also tells us how to interpret words. A machine is alleged to have translated a Russian engineer's report into English but to have included a reference, apparently zoological, to a 'water goat'. A human translator would have realised that the correct phrase was 'hydraulic ram'.

When a commonly used word has several uses, context is crucial. Hence diverse meanings, such as those of 'ram', defeat a computer. The primary meaning of *running* appears in the phrase, running for a bus. But we also have running water, a running sore, running contraband, running machinery and running a business. These are metaphors: 'running' is put in new contexts in each of which it has a different significance. Metaphors are embedded in language ('embedded' is an example). As we read, we take them in our stride (another example).

Machines cannot cope with metaphors (or puns) because they are restricted by the rules of formal logic and mathematics. These are quite different from Miss Marple's common sense. They have been systematically devised by human beings, often as an outcome of controversy. So we have a category of socially produced knowledge with tremendous power but bound by the rules—the logic—chosen by its makers. And it has to be systematically taught. (It is not taught nearly enough.) The creators are not so bound.

Are Biologists Reductionist Monsters?

The attempts to simplify human action, such as those of Pavlov and Skinner, are among the reasons why, for some people, science has a bad name. Others are its strangeness, impersonality and its uses for death and destruction. So, half way through my story of biological science, I interrupt it to ask what impression a reader is likely to have of modern biology and of biologists.

One way of finding out is to examine literature and movies. They both reflect and also help to create public opinion. In them we may find biologists represented as madmen or as blasphemers (compare chapter 2), whose derangement often takes the form of thinking only in terms of mechanisms and mathematics and of neglecting human beings. Exciting fiction often depends on such paranoid fantasies.

Some scientists in the real world do overestimate their own kinds of knowledge. Pavlov and Skinner tried to reduce the human intellect to narrowly defined performances by captive animals. Other specialists have preferred other kinds of reduction. Francis Crick has forcefully stated that biology cancels the grounds for our ethical beliefs: natural selection (the 'struggle for survival') should become the basis for building a new society.

Today, the most prominent reductionist images of humanity seem to describe us as driven, willy-nilly, by genes. The English zoologist and popular writer, Richard Dawkins, is notorious for saying that human beings are created by their genes and that we are born selfish. In a recent work, *Climbing Mount Improbable*, he describes people as self-duplicating robots. He then, in an absurd inconsistency, defines a robot as any mechanism which is set up in advance to work toward fulfilling a particular task—in effect, a machine constructed from a 'blueprint'. But neither a plant nor an animal, let alone a human being, is a machine so defined.

The reductionist writers try to explain complexities by resort to a single idea or theoretical system, such as chemistry, electronics, reflex physiology or genetics. It is easy to be impressed by the successes of this method. In the physical sciences many of the properties of the bewilderingly various materials we see around us have been partly explained by reducing them to about a hundred kinds of atom; and the atoms themselves have been reduced to ultimate particles. But we are not on that account obliged to regard all living things, including ourselves, as *nothing but* systems reducible to genes or atoms. The proposals that we should do so are, however, put forward not by the monsters of lurid fiction but by enthusiasts who do not see far enough beyond their own specialities. The reductionists themselves, like

Skinner, emerge in their own writings as human beings, often passionate and even compassionate.

The fundamental principle is this: we cannot embark successfully on reduction until we have a good knowledge of what has to be explained. We have to begin with that. Nor is this only an academic debate. When reductionist programs are incautiously applied to humanity, they can devalue us as persons: we then become preparations or mechanisms moved by forces beyond our control. Our freedom to act, our intellect and our moral principles disappear.

The present book, though written by a zoologist, assumes that human beings can make choices—a notion much debated by philosophers but almost universally taken for granted in everyday life. In practice it is acted on even by those who most strongly favour reductionist doctrines. Hence readers are held to be free to accept or to reject what I write and to be able to reach conclusions after reflection and argument. Much of the rest of the book is about units larger than molecules or even cells. It concerns the interactions of organisms (especially ourselves) with their surroundings. The remaining chapters therefore present some of the most formidable questions facing humanity. They also suggest some answers. To judge the answers and to improve on them, readers will need to make willing and strenuous use of both logic and intuition—in fact, of a whole range of human abilities.

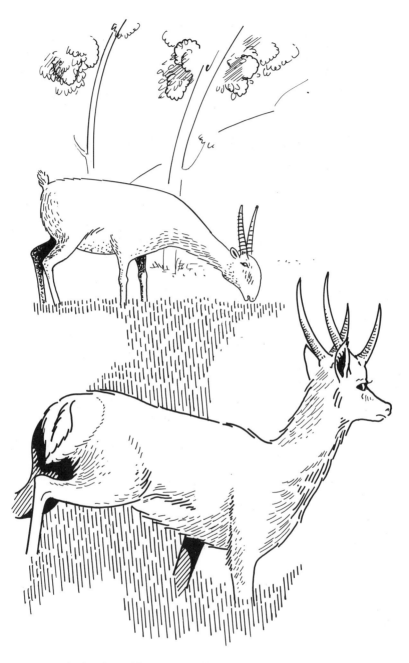

Species saved. The sheep-like creature above is an antelope, the saiga, which, despite its former millions in the Asian steppes, was nearly extinct a century ago. Below it is the pronghorn of North America which, like the saiga, was nearly shot to extinction and rescued at the last moment.

ENVIRONMENTS

The obvious biological unit is the organism. Larger units include populations and environments.

Every organism is continually acted on by its surroundings and is acting back. The environment includes members of its own kind, other species and the nonliving surroundings. Despite environmental hazards, the numbers of all species are capable of rapid growth. What stops this increase?

Wherever people live, the fate of the living things around them often depends on local, even private, action. So in the next chapter we begin with a domestic scene. Later come global phenomena, wild nature and burning issues such as plague, pollution, extinctions and population problems.

In chapter 9 we also meet new observations of animal behaviour which have led to questions (and perhaps some answers) on human societies.

Chapter 8

Ecology: Species Living Together

> Whoever could make two ears of corn or two blades of grass to grow upon a spot of ground where only one grew before would deserve better of mankind and do more essential service to his country than the whole race of politicians put together.
>
> JONATHAN SWIFT

To GET AN IDEA of the meaning of ecology, the reader is recommended to look around. You may be troubled with house mice; you therefore acquire a cat, and so use a predator for biological control of a pest. The cat, however, also attacks or drives away birds from your garden. The birds are not only a source of enjoyment: they also help to protect your flowers and vegetables from insects. Some cat owners, perhaps, live next to bushland with native species preyed upon by cats. If so, they then face yet another predicament.

Gardens

A biologist, Jennifer Owen, has illustrated a domestic situation from a small English garden, with a lawn, flower beds, shrubs, herbs, vegetables and trees. More than three hundred plant species were recorded. Variety was enhanced by species brought from afar. Among the culinary herbs, borage, marjoram, rosemary, sage and thyme were from southern Europe. Trees included sycamore from central Europe, horse chestnut from Greece, a cypress from the American West and a flowering cherry from Japan. The plants supported five species of wild mammals, 50 of birds and 55 of spiders; but most of the animals identified were of course insects, including two species never before described. (No insecticides were used.) The total of animal species listed (it was incomplete) was about 1400.

For a gardener, a crucial ecological question is what eats what. Jennifer Owen's cabbages were eaten by greenfly (aphids); they in turn were eaten by hoverflies; the hoverflies by spiders; and the spiders by birds. This is one of the many food chains observed. Gardens also show the influence of nonliving features of the environment, such as a pond or a rockery. (Leave a rock, a log or even a brick on the ground in a corner of your garden for a few weeks. Then turn it over.)

The variety was a result of careful tending. Left to itself, the garden would go though a number of stages and, after about thirty years, revert to scrub with fewer plant species. This would be an example of plant succession. Usually, however, one thinks of succession as an increase in species. The beginning of such a sequence, on the volcanic island of Krakatau, is mentioned in chapter 5. In 1883, an eruption left it a shattered surface without life. Yet a year later grasses were already growing. Other kinds of airborne life, which took part in the succession, ranged from bacterial spores to insects and spiders. A century after the explosion, the island was largely covered with rain forest.

During the 1940s, people living in European cities saw how quickly some plants can grow on bare ground. Bombs left deserted spaces among the buildings. A little soil remained, much rubble and fragments of wall. Flowering plants colonised this infertile ground, at first only small herbs but, later, shrubs and even trees. In London a botanist counted 120 species in these charming, untended little gardens.

If a garden, instead of being allowed to run wild, were converted for growing cash crops, it would support many fewer species. In regions with dense human populations, gardens are therefore a means of conserving both plants and animals; and, at present, they are not a threatened habitat: in 1991, garden lawns in the United States took up over ten million hectares of land. On one, in Princeton University, a square, 16 metres on a side, yielded 34 plant species. This type of investigation is open to amateur gardeners, children in primary or secondary school or advanced researchers. It reveals much about the diversity of species. Australian research has also revealed ancient cemeteries as valuable sites for conserving and studying biodiversity: some allow the growth of species rare or unknown elsewhere.

Soils, Recycling and Pollution

Each locality with its organisms is an ecosystem—a group of interacting organisms and their environment. On land, a large part of every system is the soil. At the surface, in the litter and just below, dead organic matter (mainly bodies, leaves and excreta) supports a vast number of decomposers and other organisms. Each species has a different appetite. Sugars, cellulose, proteins, fats and their products are biodegradable and are separately broken down by bacteria, fungi and protozoans. The end result is humus, the principal source of the substances (except carbon dioxide) needed for plant growth on land.

In the litter and humus, the microorganisms are accompanied by hosts of animals. A square metre of fertile soil may harbour a million roundworms (Nematoda), tens of thousands of mites (Acarina) and a similar number of small earthworms (Oligochaeta). These and others themselves make an ecosystem analogous to a jungle with its variety of plants, its plant eaters, its carnivores and all the rest. The earthworms eat their way through the soil and so break it up and aerate it, to the benefit both of the plants and of plant eaters such as ourselves.

Below the humus is a layer mainly of inorganic matter; below that is likely to be crumbled rock. The crumbling is partly due to plants: their roots, as they grow, force open small cracks and may eventually reduce hard stone to small fragments. In the most favourable conditions, one centimetre of soil can be formed in twenty years; in the worst, in a thousand years. But it can be destroyed in a few months.

Ecological order requires recycling. Essential elements are continually reused. During photosynthesis, carbon dioxide from air or water is combined with nitrates and other substances. These processes are balanced by respiration and decomposition, during which complex substances are broken down. Hence carbon atoms circulate incessantly in the biosphere. Similarly, as we know, nitrogen is moved around by the microorganisms which fix nitrogen and the denitrifiers which release it back to the air.

Traditional agriculture encourages the carbon and nitrogen cycles. So do gardeners. A reader with a garden probably uses a compost heap and promotes the cycles not only of carbon and nitrogen but also of elements and radicals which circulate locally.

Chinese farmers spread feces on the rice fields and so add nitrogen and other elements to the soil. In cities, however, sewers contain rivers of valuable nutrients which are often wasted or worse. When Tom Lehrer sings,

> The breakfast waste you throw in the Bay
> They drink for lunch in San José,

he is pointing not to a cycle in nature but to pollution. One remedy, increasingly used, is to manufacture a crop fertiliser from raw sewage.

Today, the world's carbon, nitrogen and other cycles are being distorted by human action. Oil bearing rocks contain enormous quantities of carbon. Of this store, increasing amounts are released into the biosphere by combustion of gasoline. One outcome is global warming. Similarly, the amount of nitrogen spread by artificial fertilisers resembles that circulated naturally by bacteria. As a result, soil fertility and crop yields have been raised, but at the cost of severe nitrate pollution of lakes and rivers.

Another cost is the release of toxins into the atmosphere. Inhabitants of large cities have long been familiar with poisoned air. Today, with the decline of open fires, we no longer have the 'pea soup' fogs for which London was once notorious. Instead we poison ourselves with blue hazes and smogs from internal combustion engines; and the gases, sulfur dioxide (SO_2) and nitrous oxide (NO_2), are producing acid rain on an international scale.

Air pollution and damage to soil go together. Acid rain produces sulfuric and nitric acids which, in soil and waters, form toxic compounds with metals. In Europe and in the east of the United States, a long period of exposure has gone with the death of millions of trees. One-third of European forests are affected. Just how the acids kill the trees has still to be found out.

Meanwhile, in some countries emission of toxic gases is being reduced. New research, however, has revealed the importance of solids in the atmosphere. Airborne dust particles, some of which are produced by industry and fires, are chemically basic and can help to neutralise the acid; they also contribute to plant nutrition. Survival of the world's remaining forests therefore requires success in a formidable task now falling on governments: to balance acid with dust.

FRESH WOODS

Each major climate has trees of a typical structure. On tropical coasts mangrove trees are the most conspicuous. Mangroves are very odd: they thrive in sea water, some by excreting salt from their tissues. The swamps where they grow have little oxygen, but the trees have strangely modified roots which project into the air and allow respiratory gaseous exchange. Above high water, mangroves provide a canopy in which are great numbers of wood boring beetles and moths; the burrows are used by other insects and by mites and scorpions. In the next zone down the stilt roots of the trees carry red algae and a menagerie which includes crabs, oysters and barnacles. Many species are not found elsewhere. Also present is a very queer fish: *Rivulus marmoratus* is a hermaphrodite said to be able to reproduce by self fertilisation. Lower still is a zone where mobile oysters and crabs thrive among sessile sponges and anemones on a matting of sea grass.

Mangrove forests are not well studied. Many of the species they shelter have still to be identified. A systematist who wishes to add to a record of new species discovered and described should hasten to collect them. For these fragile ecosystems are being steadily destroyed. Few shores are unaffected by human action. The swarming human populations are often especially dense on the coasts. The life on rocks, in sand or in mangrove swamps and other associations, then declines or disappears. Even where shores are rarely visited by people, the seas which wash them are increasingly polluted by industrial wastes, sewage and plastic.

Far from the mangroves, in the cold north of Asia, Europe and America, pines and other cone-bearers form a broad zone round the earth. Many have to cope with cold as low as −40°C, heavy snow and yet shortage of water. Their leaves (pine needles) are too thin for much snow to settle on them; they are dark and absorb much heat from the sun; they lose water only slowly; and the little sap they contain does not readily freeze. The pyramidal shape of a pine tree prevents any great build-up of snow on the branches.

In temperate climates most trees have thin, broad leaves which take maximum advantage of the sun and allow passage of much water: a large oak in summer transpires many tonnes of water in a day and rapidly manufactures sugars and other substances. In yet warmer climates, summer is the most exacting season. Many trees, notably the Australian 'blue gums' (eucalypts), have narrow drooping leaves through which water passes slowly. Some eucalypts, however, grow quickly and have been planted on a large scale in Southern Asia and North America for timber.

In the absence of human interference or other disaster, a forest may appear a stable ecosystem in which, century after century, input balances output. The expression, the 'balance of nature', expresses this idea. It can, however, be misleading. Fire may be a calamity, but for some forest systems it is a necessity. Without fire, shrubs may form a dense ground cover. Eventually, lightning or some other natural event then starts a disastrous conflagration. High densities of the giant sequoias of California depend on fire. The thick bark of these superb trees is exceptionally resistant and the seeds germinate best on ground left bare but with plenty of minerals. In large areas of American and other forests, fire is systematically used as part of management.

Well within historical times, much of Europe and Britain were heavily forested. Over centuries, the trees were largely replaced by plowed fields or pastures. The result of this longterm management was a slow conversion to a stable system of agriculture and stockbreeding. Elsewhere, forests, especially in South America, South East Asia and Australia, tell another story. Tropical forests differ from temperate woodlands in the poverty of their soils. Root systems are shallow. Recycling of substances occurs in the surface litter above the soil and in the thick tangle among the living plants and animals. Small plants (epiphytes) which grow on large ones play an important part;

so do the lichens and algae which are also on the trunks and branches of trees. Successions are always in progress. When a large tree falls it leaves a clearing. Sun-loving herbs and grasses arrive. These fugitive species appear as opportunity offers. Then come trees which grow quickly. The original large tree is replaced only after many years.

A region of rain forest is not a single ecosystem but many. In the largest, that of the Amazon basin, some areas are regularly flooded by the enormous rivers flowing through them. The rivers themselves vary. Some carry clear water and support the tallest trees and the forests with the greatest variety of species. Others are muddy or black with dissolved plant matter; these too supply distinctive associations of plants and animals. Most of the species are found nowhere else: they include almost blind dolphins (*Inia geoffrensis*) which find their way among the tree roots by echolocation (sonar); a great variety of birds; caimans (related to crocodiles); turtles; and (it is estimated) about 3000 species of fish, many of which have modified teeth and feed on fruit and seeds. The uncounted people who live on the river banks depend largely on the fish for food. The turtles include a large species, *Podocnemis expansa*, which may weigh 50 kilograms; the females lay more than a hundred eggs at a time; the young feed on plants and grow rapidly. Properly managed, the turtles could become a substantial source of animal protein.

Among those who have achieved equilibrium with the living environment are the rubber tappers who have for long milked trees without destroying the ecosystem. But today tropical forests of all kinds are disappearing at an increasing rate. The causes are many. At one extreme are the poor cultivators who cut down or burn trees and shrubs in order to survive. At the other are loggers, who devastate large areas for timber, and the stockbreeders who fire great tracts of forest and sow them, during the rains, with grass seed scattered from aircraft. After further firing, the cleared area, now grassy, is ready for cattle. But in a few years it becomes useless scrub which has to be abandoned: the soil and the climate in such regions are suitable for pasture, at best, only with careful management.

Longterm conservation demands more than restrained farming. Converting established forests to farmlands, even if done responsibly, carries ecological costs which have only lately been understood. Among them is habitat fragmentation: a large region is often broken up into a patchwork. Portions of the original vegetation may remain, but they are likely to be too small for the survival of many species which thrived in the original forest. (In temperate regions this has most obviously affected large predators, such as wolves, bears and eagles.)

How can the remaining tropical forests be preserved and, at the same time, be used for human benefit? A recent study of the upper Amazon region has revealed a wealth of forest resources available from the growing trees: not only chocolate and rubber but also many vegetables and fruits. The planned use of these products, combined with strictly regulated logging, has proved to be more profitable, over a few years, than ruthless clearing.

PASTURES NEW

About a quarter of the world's land surface is covered with grasses. Their zone has a rainfall between those of forests and deserts. The flowers of grasses are unobtrusive and do not attract insects: the pollen and seeds are carrried by the wind over the open plains. The leaves grow continuously at the base: cropped by grazing animals they grow up again. The roots form a continuous tangle which holds the soil together. Many grasses also have runners which sprout new leaves.

As a result, grasslands can survive both drought and torrential rains. But they do not always do so. During the 1930s, large areas of North American prairie suffered a disaster. After plowing, wheat (which is a grass) had been sown, year after year, without precautions against loss of soil. This, and a low rainfall, resulted in the notorious dustbowl: the fertile soil, a product of millennia of natural recycling of matter, dried, crumbled and blew away. Much of what remained was desert. One positive outcome was a new public interest in conserving the environment. Another was a program of repair: among many measures, new species of plants were introduced to bind the shifting sands; on sloping ground, contour plowing and terracing were used to prevent the remaining soil from being washed away.

Stable grassland has a rich soil and supports a great variety of animals as well as plants. The most imposing animals are the large mammals of which, until recently, enormous numbers thrived in Asia, Africa and North America. Such herds still flourish in East Africa. There the savanna, though drier than most of the major grasslands, can support stable food chains or networks. Termites and ants help to crop the grass and contribute to soil formation. The insects are consumed by strange mammals, the aardvark (*Orycteropus afer*) and the pangolin (*Manis*), as well as many birds. The soil is burrowed by rodents which eat the plants and are eaten by jackals (*Canis*), weasels (*Mustela nivalis*) and others. Large carnivores, including lions (*Panthera leo*) and hunting dogs (*Lycaon pictus*), prey on the magnificent collection of the largest herbivores: elegant, agile gazelle; oryx and eland (*Taurotragus oryx*); wildebeest or gnu (*Connochaetes*); giraffes; and others.

Each of these has to cope with the tough plants, shortage of water and the intense heat of the African day. The large, rather bovine eland has been studied by telemetry: the output of miniature broadcasting devices, implanted in the animals, can be detected at a distance. An eland keeps cool partly by standing in the shade, partly by panting, but it can put up with a substantial rise in temperature—a high fever. (Domestic cattle of temperate lands cannot tolerate raised body temperatures: they keep cool by sweating, a method which squanders water and is unsuitable for Africa.) Elands can be a large source of human food. So can oryx, which feed on diverse plants at night and from them get all the water they need.

On the Asian steppes is the saiga (*Saiga tatarica*), an antelope which looks rather like a sheep. Until the nineteenth century, it lived in herds each of tens of thousands of animals. By 1900, Russian guns had reduced it to apparent extinction. At the last moment, shooting stopped and preservation began. In half a century the saiga recovered from a few hundred to more than two million. It is now a source of meat.

The story of the bison on the North American prairies resembles that of the saiga without the recovery. The male is taller than a man and may weigh 1000 kilograms. Early in the nineteenth century these formidable animals may have numbered more than thirty million. The scattered bands of Plains Indians acquired their food, clothing, tools and cups from thriving herds. By 1900, systematic slaughter by Europeans had left only a few hundred. Today, some thousands are protected in national parks.

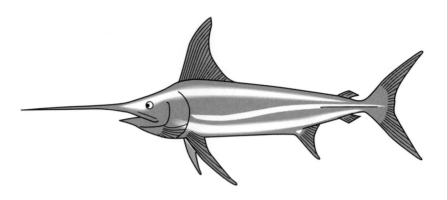

A very fast fish, the swordfish, *Xiphias gladius*, which, like its close relatives the tunnies, lives in open seas.

Most large grass eating mammals are ruminants: they include deer, giraffes, sheep, goats and cattle. All have complex stomachs and digestive processes. Grass first goes into a chamber of the stomach where it is partly digested with the aid of a symbiotic population of bacteria and protozoans. The resulting cud is returned to the mouth for chewing; finally it is further processed in other parts of the stomach.

One group of large herbivores manages, however, without ruminating. The horses (*Equus*) began in North America, became extinct there, but spread through Africa and Asia. Africa still has great herds. There the horses are striped and cannot be tamed: they are zebras. The domesticated species (*E. caballus*), a zebra without stripes, exists in its original wild state only in Central Asia. But it is hardly a threatened species.

ALL AT SEA: MADHOUSE ECONOMICS

We now go to sea. Seventy per cent of the earth's surface is water. In the middle of the Atlantic Ocean, a continuous whirlpool drives floating objects into its centre. There, as Christopher Columbus (1451?–1506) and other mariners found, the surface is thick with a brown seaweed (*Sargassum*). Today, the same region also accumulates plastic shopping bags.

The pollution of the Sargasso Sea is an index of what is happening to marine life. The seas support beautiful and edible animals whose numbers have declined, some almost to extinction, during the twentieth century. Among them are the tunnies and their relatives of the mackerel family (Scombridae). The tunnies represent the peak of design for speed in surface waters. The largest, a swordfish, *Xiphias gladius*, whose snout ends in a spike, can reach a length of six metres. The fastest, a sail fish, *Istiophorus*, has been timed at 110 kilometres an hour over short distances. The internal temperature of most fish is the same as that of the water, but the muscular exertions of these high speed predators can raise their temperature well above that of their surroundings.

They, and their smaller relative, the mackerel (*Scomber*), head a food chain which begins with the microscopic floating plants; this phytoplankton is eaten by microscopic crustaceans (members of the zooplankton) which, in turn, are food for fish of modest size, such as herring and sardines; these crowd together in shoals and are eaten by mackerel and their larger relatives and by people.

The powerful muscles of tunnies are highly edible. The intensity of fishing them and other edible species has gone up steeply with the increase in human numbers. Early in the twentieth century, daily catches in the North Atlantic began to decline. Overfishing was suspected, but little was known about fish populations. The two world wars later provided some evidence: fishing was interrupted during both; and, after each, at first larger fish were caught and fishing was more profitable than before war broke out. From about 1950, profitability was further increased by improved techniques and the use of ships with advanced methods of navigation and novel equipment: enormous, tough but light nets of plastic; machines to haul in the nets; spotter aircraft and electronic means of locating shoals; and refrigeration. From the late 1940s to 1988, the world's annual catch of fish rose from about 20 million tonnes to 85 million. By 1970, however, only half the catch was still being eaten by people directly: the rest was processed as margarine or as fish meal for domestic animals.

More recently, catches have steadily declined and much of the new equipment has gone out of use. The formerly gigantic populations of cod (*Gadus callarius*), mackerel (*Scomber scombrus*) and herring (*Clupeus harengus*) in the North Atlantic have collapsed. Catches in the Pacific and the Mediterranean have gone steeply down. Fishing fleets are stranded or kept at sea by state subsidies. The world's fishing industry has been described as being run on the economics of the madhouse. During attempts to prevent worse overfishing, clashes have occurred between fishing fleets and gunboats of North Atlantic nations. Is this strife a portent of a future of declining resources and rising human numbers?

For some fisheries, however, the latest news is better. During 1900 to 1940 the catch of pacific salmon (*Salmo nerka* and others) rose to a peak; but it then fell: by the 1970s it was only about ten per cent of the maximum. At that time, programs of conservation were begun in Alaska and Canada, based on knowledge of the biology of the several species of salmon. Salmon spawn in rivers but spend much of their lives in the sea, where they are netted. Conditions in the rivers have been improved; artificial spawning channels have been provided; hatcheries have been set up; and fishing has been restricted. As a result, in the 1980s, populations and catches had been substantially restored.

The yield from other fisheries similarly requires global research and planning. Accurate prediction is therefore needed. When the population of an important food fish, such as cod, is left to itself, its numbers fluctuate from year to year, often for unknown reasons. Changes also occur in the age structure of each population: at one time few large adults may be present but, at another, many adults and few young. For maximum sustainable yield the changes must be monitored. The intensity of fishing can then be adjusted by a central authority.

An alternative to hunting fish is aquaculture in large tanks or in artificial ponds or lakes. Carp (Cyprinidae) of several species are very productive. A tropical freshwater genus, *Tilapia*, which has many species and has been severely overfished in lakes and rivers, can do even better than carp.

The largest living mammals, the rorquals or fin whales, have a recent history similar to that of edible fish. They are toothless and feed on small, shrimp-like creatures, krill (*Euphausia*), which thrive in vast numbers in surface waters. The blue whale (*Balaenoptera musculus*) may be more than 30 metres long and weigh over 100 tonnes. Even at birth it is about eight metres but, supported by water, it has no weight-bearing problems.

Whales secrete a very concentrated milk. A writer of science fiction, A.C. Clarke, has described, in *The Deep Range*, a near future in which toothless whales are treated as dairy cattle. Certainly, the yield of rich, creamy milk would be impressive.

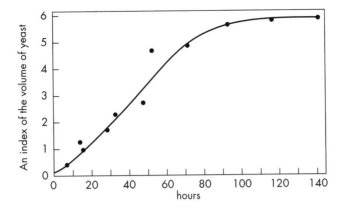

A limit to growth. The growth of a population of yeasts in the laboratory. When yeasts are few, increase is slow; later, the growth rate accelerates; but eventually a density-related factor slows and stops it: in this case, the concentration of the alcohol produced by the yeasts.

One group of cetaceans, including the dolphins (Delphinidae), have teeth and are predators. Dolphins get on very well with human beings. So Clarke also suggests that toothed whales could be trained as counterparts of sheep dogs.

As far as I know, research on milking or herding real whales has not yet begun. Instead, some species of whales are threatened with extinction. The first major hunting nearly killed off the right whale (*Balaena glacialis*)—called 'right' because it is easy to catch; it remains rare. In 1868, the harpoon gun and explosive harpoon were invented and thereafter the gigantic blue whale was relentlessly pursued. When hunting blue whales was no longer profitable, the common fin whale (*Balaenoptera physalus*) was similarly almost wiped out. Late in the twentieth century, however, international agreements are slowly lessening the threat to whales. If, for a quarter of a century, whaling stopped or almost stopped, these magnificent animals could again become abundant.

BEAUTIFUL CURVES

Most organisms can in fact breed much more quickly than is needed for constant numbers: reduce a population and then leave it and it soon bounces back. But eventually it stops. It is not always at all clear what stops it.

A microorganism, such as a yeast or a protozoan, can be

cultured in a regularly replenished nourishing fluid in a laboratory. The numbers may at first be very small. (They do *not* begin at zero. Some textbook diagrams imply that a population can begin by spontaneous generation from sugared water.) Increase is slow at first, then more rapid; later, as in the graph, the curve flattens and the population settles at a constant maximum.

Does this simple, S-shaped curve tell us what happens outside the computer or the laboratory? If it does, seemingly most populations ought to be steady at the top of the curve; yet in fact nearly all fluctuate, if only because the weather does so; they also interact with other species which, too, are varying. This difficulty, as we know, arises with fish populations and is important economically.

Some populations seem to do nothing but vary wildly. Two Australian ecologists recorded the numbers of an unobtrusive insect nearly every day for fourteen years. The insects were thrips (*Thrips imaginis*), each about one millimetre long. They live in the flowers of many plants. The average number in a rose flower ranged from about 100 to 500. Even at the maxima, no evidence was found of overcrowding or shortage of food: the weather seemed to be the controlling factor. The numbers of many other species, especially of insects, are even less predictable. This applies to some important pests, notably locusts, of which more below.

The slowing down and final ending of growth suggests some adverse influence, of which the effect *increases with population density*. For the yeasts shown in the graph above, such an influence is easily found: the cells stop dividing when the alcohol they produce reaches a certain concentration. Here is an example of a negative feedback acting in a population. (Compare page 39 on negative feedbacks in cells.) If we could find such density-related factors in the outside world, we should learn much about how populations are regulated.

Possible checks on animal numbers

Food or water lack
Shelter shortage
Predators
Pathogens: microorganisms and parasites
Social interactions

Food, Shelter and Predators

Above are possible density-related factors for animals. The most obvious are food and water. Correspondingly, when we want free-living animals to multiply, one method is to encourage the growth of food plants. This, however, can lead to complications. Great efforts have been made to improve conditions for black-tailed deer (*Odocoileus hemionus*) in the United States. These attractive animals can breed at a high rate but, left to themselves, they feed selectively on the most nutritious plants and are likely to eat too much. They then convert a favourable range to a poor one and their numbers decline. For management, they have to be regularly culled as well as fed.

A population may exhaust its shelter before it runs out of food. The pied flycatcher (*Muscicapa hypoleuca*), a charming little bird of European woods, feeds on insects and nests in holes in tree trunks. In a wood in Finland with only a few of these birds, many nest boxes were installed. The boxes were quickly occupied and the numbers of flycatchers rose. Gardeners who install nest boxes can similarly change the face of nature in their neighbourhood.

When we come to predation, it is often difficult to decide whether the predators are controlling the prey or the prey are determining the numbers of predators. When an English wood was studied for thirteen years (not a long time in ecology), the number of tawny owls (*Strix aluco*) seemed to depend on the numbers of the voles and field mice on which the owls fed.

Occasionally we find a clear case of predators controlling prey. The red kangaroo (*Macropus rufus*) in Australia was helped by the European invasion and the consequent increase in grassland. Kangaroos are preyed upon by dingos which also attack sheep. In the south-east of the country shooting and poisoning has almost wiped out the dingos; and a fence, 9660 kilometres long, keeps them from the cleared area. In 1980, inside the fence, in the region without dingos, red kangaroos were numerous (and, sometimes, pests); outside, they were eaten by dingos and were rare.

Predators can be introduced to attack pests. Late in the nineteenth century, a bug, the cottony cushion scale (*Icerya purchasi*), infested orange and lemon trees in southern California and nearly destroyed the citrus industry. The home of this insect is Australia, where it is preyed upon by a ladybird beetle, *Rodolia cardinalis*. In 1898, a few of the beetles were brought over, at the trivial cost of $1500, and released near Los Angeles. After only a year the Californian oranges and lemons were almost free of the bugs—a notable example of biological control of a pest. Just enough of the bugs remained to allow the ladybirds to survive in small numbers: predator and prey were in equilibrium. So California remained virtually free until, long afterwards, in some areas, DDT was rashly used against other insects. Most of the *Rodolia* were killed and the scale bugs temporarily returned.

A major advantage of biological control is that the predators usually attack only the target species. They are quite unlike poisons which kill insects indiscriminately—and, often, not only insects. Pest populations can also quickly become resistant to insecticides, but they do not so readily adapt to the attacks of predators. In addition, the method leads to great savings for producers. To make use of our many potential allies, we therefore need to know in detail the life histories of the pests and of their enemies. During 1888 through 1972, the United States government spent $20 million on such research; whereas, in *one* year at the end of that period, the US chemical industry spent more than five times as much on the profitable business of developing synthetic pesticides.

RABBITS AND MYXOMATOSIS

Parasites, including infective bacteria and viruses, are, like predators, possible regulators of numbers. Their action is related to density, because the chance of infection goes up as a population becomes more crowded. Most organisms which live in or on other organisms are, however, in equilibrium with their hosts: even if not symbiotic they are at least harmless. This rule is exemplified by the myxoma virus. Yet, paradoxically, the virus also provides the most famous successful use of a disease agent against a pest.

In Brazil, some of the wild rabbits or tapetis (*Sylvilagus brasiliensis*) are infected with myxoma without suffering much harm. When, however, European rabbits (*Oryctolagus cuniculus*) were infected the result was usually myxomatosis, which was nearly always fatal after about two weeks. This finding was applied where rabbits were at their most destructive. The

European rabbit had been released in Australia by English colonists who proposed to trap or shoot it for food. In the absence of effective predators, it multiplied and spread at an astonishing rate. Its fertility, and its destruction of pastures, became so horrifying that it seems to have induced hallucinations. Some stockbreeders in dry regions are said to have reported a new variety of rabbit which had a hump and could go for days without water.

In 1950, wild rabbits in Australia were believed to number about 500 million. In that year, in the south-east of Australia, after many attempts, they were successfully infected with the myxoma virus. It spread rapidly, carried by mosquitos. The insects themselves are passive carriers, like the hypodermic needles which carry AIDS. In one year about 80 per cent of the rabbits in a large region were killed. Soon the rabbit population of the whole continent had dwindled. The remaining rabbits are, however, less susceptible to the infection—a result of selection for resistance to myxoma. Less virulent strains of the virus have also arisen: these can immunise rabbits against the virulent form. The rabbits have therefore begun to return. Lately, however, a new infection, viral hemorrhagic disease (calicivirus), has been released and has produced another (temporary?) decline in rabbit numbers.

Myxomatosis has also infected the rabbits of Europe and Britain. When a widespread and numerous species is largely wiped out, side effects must be expected. In Britain and elsewhere, predators such as foxes and eagles, deprived of rabbits, are said to have turned to hunting other prey, including poultry and lambs. A desirable outcome has been the appearance of flowering herbs in large numbers, where rabbits had cropped grasslands so closely that few flowers could grow.

Social interactions, the last of the density-related factors, belong in the next chapter. So here are two more histories which tell us more about the intricacies of ecology. Both include population explosions and have been of vast social importance.

Locusts

Much of Africa, the Middle East and India are subject to devastation by intermittent locust swarms. In the Americas and

Australia, related species assemble in great numbers on the ground. In 1957, a plague of the desert locust (*Schistocerca gregaria*) darkened the land of Somalia. The number of insects was estimated as 160 000 million and their weight as 50 000 tonnes. The insects were so closely packed that, when they settled, they sometimes brought down large trees. Each locust ate about its own weight of food every day. Similar calamitous, unpredicted and unexplained attacks have been recorded for millennia.

Until 1921, the biology of the eight locust species was an enigma. Locusts are grasshoppers (family Acrididae) but they were known only when they were swarming. Where were they at other times? The fundamental discovery was made by a Russian entomologist, B.P. Uvarov, who later made his headquarters in England. Locusts, when not swarming, live spread out on the ground. The migratory insects of each species differ sharply from the solitaries. The solitary forms had therefore already been classified and named as separate species. Solitary individuals of an African locust, *Locusta migratoria*, are light green, have small heads and fly at night. The migratory forms crowd together, are strikingly coloured black and yellow, have large heads and fly during the day; they also have extra internal food reserves.

In a further crucial discovery, certain regions have been identified as outbreak areas. They have a dry climate but sometimes exceptional rains allow extra growth of plants. Numbers then rise. Evidently, most of the time, food limits numbers. The insects crowd together and move off in groups, at first on foot. Individuals of the migratory type appear. Finally, millions become airborne. One swarm, closely observed, began in East Africa and was carried more than 3000 kilometres to Jordan and Iraq.

Watchers in the outbreak areas can now predict outbreaks. Before they take off, swarms can then be sprayed with insecticide from aircraft. In the 1950s, as a result, for the first time in history Africa survived a period of locust attack without serious loss.

The locust story represents a major achievement in both academic and economic biology. Effective action has depended on the efforts of many specialists: entomologists had classified the hundreds of species of grasshoppers; ethologists had experimented on their behaviour; physiologists had recorded changes in locust metabolism; ecologists had studied plant and animal populations in the harsh conditions of African grasslands. In addition, predicting outbreaks requires the unremitting labours of many field workers; and preventing swarms requires insecticides and spreading the chemicals in the right places. All these activities have to be financed by governments. Finally, full success requires the cooperation of many nations which are not always in cordial agreement.

Plague

Like locust swarms (until recently), plague arrives in human communities without reliable warning. For millennia it has brought panic, death and disruption in Asia, Africa and Europe. Eventually, it also entered the Americas. It strikes at long intervals, continues for decades or even centuries and then fades away. Its ending has been as obscure as its origin. In sixth century Europe, the plague of Justinian killed uncounted millions. Another pandemic began in the interior of Asia in about 1333. This, the Black Death, travelled through India, reached Europe by sea in 1346 and soon spread over nearly the whole continent. Death rates in small communities, such as towns or villages, were sometimes 75 per cent. The last flaring of the disease in England was the Great Plague of London in 1665; France had an outbreak as late as 1720 in Marseilles. We are now surviving another pandemic. This too began in Asia, near the end of the nineteenth century. In India, during the first decade of our century, deaths from plague were more than a million annually. The disease spread from India throughout much of the world. South Africa and the west of the United States were among the regions affected.

In an outbreak, the proportion of infected people is rarely less than ten per cent and is usually much higher. Lymph nodes in the groin, armpit or neck enlarge and form swellings or bubos (hence bubonic plague); a high fever develops; black patches appear under the skin (hence Black Death); about half of those infected die in a few days. Sometimes the lungs are involved (pneumonic plague); people so infected have about a day to live.

Until the 1890s, no authentic treatment or means of prevention existed. Plague was a source of terror variously attributed to the 'cruelty of Heaven', to 'the devil's work' or to an unfortunate conjunction of the planets. Belief in heavenly causes of plague became obviously futile only when, in 1894, a Japanese bacteriologist in Hong Kong, S. Kitasato (1852–1931), discovered the bacillus responsible. At the same time, in France, G.A.E. Yersin (1863–1943), a pupil of Pasteur, also discovered the bacillus (now called *Yersinia pestis*) and developed an antiplague serum.

The plague organism has a permanent home in many species of rodents, including voles (*Microtus*), gerbils (Gerbillinae) and marmots (*Marmota*). Many live in forests, especially those of Mongolia and Siberia; others live in grasslands. Some carry the infection without ill effects. (Compare myxoma in South American rabbits.) Occasionally, trappers and others working in these sparsely inhabited regions are infected (sylvatic plague). The 'sylvan' environments include the wilder parts of California, where deermice (*Peromyscus*) make a reservoir of infection. The bacilli are carried among all these mammals by blood-sucking fleas. They become a danger to large numbers of human beings when rodents which live among people also become infected. The most important are rats, especially the 'black' or 'roof' rat (*Rattus rattus*), but also the 'brown' or 'Norway' rat (*R. norvegicus*). Guineapigs (*Cavia*) too can carry the disease. Danger is greatest when rats in human communities die of plague and their infected fleas resort to the tougher human skin.

The natural history of this one organism, the plague bacillus, therefore involves an interaction between many rodent species, their fleas and the human species. The numbers of the rodents fluctuate; the bacillus itself changes in virulence; and the hosts, both rodents and human beings, vary in their ability to resist infection. Prevention is by restricting the movements of rats, especially in ships; by killing rats and their fleas; by denying rats the opportunity to breed; and by immunising people. Treatment is by antibiotics.

Twentieth century methods can be applied against plague (as against AIDS) only by communities with well organised health services. Social upheavals allow the reappearance of epidemic plague. In the 1960s, during the war in Vietnam, at least 5000 people died of the disease. In 1994, in the Indian State of Maharashtra, an outbreak led to many deaths. Civil disturbance also resulted. The authorities had been warned of danger from increasing numbers of dying rats, from other infected rodents and from squalor. Yet the state government, bent on reducing public spending, had closed the organisation that watched for plague. We cannot rid the world of the bacillus. If it is not to devastate whole communities, continuous vigilance and well endowed departments of public health are needed.

EXTINCTIONS AND INVASIONS

If the plague bacillus were completely destroyed, few would mourn it. Human beings are unremittingly in the business of destroying species: the loss of those that cause disease is welcome.

What of organisms, the majority, that do us no harm? In one aspect, extinctions are 'natural': the continual appearance of new forms during evolution has gone with a similar loss of others. In the remote past, mass extinctions seem to have occurred rather abruptly. The most notorious is the one supposed to have wiped out the dinosaurs. We can, however, confidently explain only those for which humanity is responsible. We are now in the early stages of a mass extinction of living forms, similar in scale to those of the past but much more rapid. The losses have often been due to hunting or to introducing foreign organisms; but some whole environments have been disrupted.

Many species have been hunted to death with only the simplest equipment. In New Zealand, Maoris found large flightless birds, the moas, which were edible and easily caught. After a few centuries, none remained. Large mammals have proved to be similarly vulnerable: great exertions are now needed to preserve elephants, rhinos, whales and others.

In many regions, newly introduced organisms have caused disaster. Among them, the house mouse in Australian farmlands intermittently breaks out into vast, destructive 'plagues'. Other introductions, like that of the rabbit, were deliberate but misguided. The fox (*Vulpes vulpes*) is an inedible species which can, however, be hunted ('the unspeakable in full pursuit of the uneatable'). In Australia the distinctive fauna of small mammals is still being steadily eroded; and the fox, introduced for sport, aided by domestic cats, is making its contribution.

Some highly successful introductions have carried an unexpected cost: in North America, Australia, and elsewhere goats, pigs, cattle and horses have gone wild and done much damage. In Australia, wild horses or brumbies have been so prolific that they have been shot as pests or have been killed for their skins. Pigs are often worse: in California their diet has included peaches and clams.

New introductions still occur. In France, in the 1990s, a comic but alarming example has arisen from the craze for 'ninja' turtles. Small terrapins (*Chelydra*) from North America have been bought as pets for children. Many have escaped or been set free. Some have grown large and have taken to preying on fish and even on people bathing.

Disastrous invasions include those of weeds, notably in the United States. When New England was colonised, the newcomers included couch grass (*Agropyrum repens*), dandelion (*Taraxacum officinale*) and stinging nettles (*Urtica*). By the middle of the nineteenth century, on the other side of the continent, more than ninety species of foreign weeds were doing well. The grasslands of California were among those largely destroyed by invaders such as thistles (*Carduus*), ryegrass (*Lilium perenne*), wild oats (*Avena fatua*) and black mustard (*Brassica nigra*). Plantain (*Plantago*), one of the sacred herbs of the Anglo-Saxons, was called Englishman's foot by the North American indigenes and flourished as far south as Argentina. A single plant may produce 15 000 seeds of which more than half germinate. The leaves spread out and exclude other plants. An underground system allows survival during frost or casual weeding.

Among the whole environments destroyed, the most obvious are the forests. Recently, in Ecuador, a large ridge, covered with trees, was surveyed by botanists and found to contain many new species. Immediately afterwards, the trees were destroyed by loggers and the newly found species with it. No zoological survey had been made. We shall therefore never know what animals were also lost for ever.

Such losses of biodiversity, due to human action, are going on all the time, in every major type of ecosystem. The case of Lake Victoria is notorious. This mass of fresh water, the size of Switzerland, borders four East African countries. Until recently it was the home of more than 300 species of fish of the family Cichlidae. (Colourful members of the family, such as *Haplochromis*, are familiar in aquaria.) These fish, though small, made a large part of the food of the shore-dwelling people. In the 1960s a new species of fish was introduced, the Nile perch (*Lates niloticus*), which can grow to the size of a small man. The perch flourished and became a source of both food and trade. They also ate the cichlids, of which more than half the species have been wiped out. The population of perch is now collapsing for want of food and the ecology of this giant lake has been irretrievably disrupted.

When anxiety is expressed about the disappearance of species, it is sometimes argued that, since extinctions are characteristic of life, what we are observing is natural and inevitable. But, as the fate of many fish, birds and mammals shows, the present changes are far more rapid than those of the past. Out of 9000 modern species of birds, 20 per cent are endangered or already extinct. Every year, one or two species are lost. About some species of mammals, however, we may have mixed feelings. The wolf (*Canis lupus*), though a fascinating and endangered animal, would not be welcomed back to the remaining wooded areas of northern Europe or to the New Forest in the south of England; but, in North America, it is protected in some nature reserves. Even the attempts to preserve the tiger (*Panthera tigris*), another beautiful and threatened species, are likely to be looked at askance by Indian villagers endangered by a man-eater.

Despite such quandaries, what has been called the absolute obligation to preserve biodiversity for future generations is increasingly recognised. Eighty countries have now signed a Convention on International Trade in Endangered Species. As a result, trade in several hundred rare forms is illegal. To the extent that this agreement is enforced, it will help to preserve some splendid and fascinating creatures. If it fails, our descendants will be deprived of still more species for ever. But, at best, its effects will be small: it will not protect habitats. In the first decades of the twenty-first century, further unwelcome losses can hardly be avoided. We return to the dangers to the biosphere, and to ways of avoiding them, in chapter 12.

HUMAN NUMBERS

The current destruction of species and habitats has gone with the recent steep increase in human numbers. During nearly all the 100 000 years or more of our species' existence, our ancestors were wanderers, without permanent settlements or agriculture. A visitor from space would have found *Homo sapiens* a rather rare species. The number of people in the whole world may have been between four and ten million. The larger figure is less than the present population of New York City or Greater London.

Our ancestors were, however, impressively versatile: they occupied every continent except Antarctica; the environments of people without agriculture have included tropical forests, deserts and the Canadian Arctic.

Something can be surmised about their populations, especially from what we know of the living pre-agricultural peoples of the Kalahari in South Africa and of Australia. Gatherer hunters are sometimes thought to be continuously short of food; but we have no good evidence of chronic food shortage either from those that survive or from the traces of past populations. Modern gatherer hunters have, however, a high rate of miscarriage and, more important, their fertility is reduced by the custom of breast feeding each child for several years. While a woman is giving milk on demand, day and night, she is usually infertile. As a result, she may have only three or four children in her lifetime, instead of the possible twelve or more. Probably, before agriculture, human populations were kept sparse by low fertility and the dangers of a wandering existence.

Prolonged breast feeding, however, is not an automatic, unavoidable regulator of numbers: unlike the density-related agents which control animal populations, it is optional. If other food is available, a child can survive when weaned from the breast after only a few weeks. Whether breast feeding for three years is adopted depends on decisions made by women; these in turn depend greatly on local tradition.

When human beings began to cultivate food plants, human numbers rose. As far as we know, this first happened, more than 10 000 years ago, in the productive high ground between the eastern Mediterranean and India—the 'fertile crescent'. The people of this region had long been eating the seeds of grasses. Some collected them and ground them into flour. Eventually, cultivation began to replace gathering; and gradually, over many centuries, farming became important. Similarly, wild goats, sheep, cattle and pigs had been hunted. Now they began to be herded into the new human settlements and encouraged to breed there. Eventually, quite small areas could produce enough food for a human population which before would have needed many hectares.

So human life became less precarious. One consequence was harder work: farming demanded much more than the gatherer hunter's two or three hours a day getting food. Another outcome may have been an inferior diet. Evidence from Africa suggests a decline in stature and bone growth with the adoption of agriculture. Yet fertility evidently rose and allowed a steady increase in human numbers. Whether death rates went down is doubtful.

A mysterious fact is the rise of agriculture in South East Asia and Central America at perhaps about the time of its appearance in the Middle East. Each region has its own distinctive kinds of cultivated food plants. Did farming develop in one place and then spread, or was it independently invented in three widely separated regions? Most archeologists believe the latter. Whatever the answer, in a few thousand years the environment of most human populations was drastically altered. That transformation must have been a result of human planning. For more than two million years, human beings and their predecessors had been making stone tools. A toolmaker must imagine in advance the shape of the future hand ax or other item. Now, as they sowed seeds as well as eating them, the pioneer cultivators had to imagine future crops.

Most gatherer hunter groups probably included fewer than a hundred people. The earliest farming communities sometimes had several hundreds. Once agriculture began, the world's population rose slowly until modern times, partly by spreading into new regions. The time it took to double was about 1500 years.

What, in these conditions, were the restraints on numbers? In historical times, famine has occurred intermittently but persistently. Exceptional weather could at any time result in failure of crops and starvation. But, probably more important, the denser human populations provided a favourable environment for disease organisms. Parasites had better opportunities to move from one host to another. Domestic animals were sources of many infections. Human settlements were not hygienic: excreta accumulated. Drinking water was a source of killers such as typhoid and cholera—and often still is. Mosquitos bred in pools and streams and transmitted the worst endemic disease of all, malaria. One consequence was a high death rate among children.

Nonetheless, two centuries ago, human numbers took off. During the nineteenth and twentieth centuries, the most serious infectious diseases have, in many countries, been overcome. Partly as a result, death rates have declined steeply: in the richest countries it is usual to live to over seventy; and, even in those with large peasant populations, such as India, the expectation of life has greatly risen.

Near the end of the twentieth century, the total human population reached about eight times its probable size in 1600. At the end of July 1990, China reported a population of 1 133 682 501. No doubt the seven million census takers made a few mistakes; and all the figures in the previous paragraphs are estimates subject to quite large error; but, certainly, more Chinese are alive now than there were people in 1800.

The rate of increase in human numbers is, however, declining. In some industrial countries the birth rate only just exceeds mortality. Such countries have undergone the demographic transition: the decline in death rates has been followed by an even greater decline in birth rates. The causes of the lowered fertility evidently include unannounced decisions, by many couples, to restrict family size by whatever methods are available.

Even in poor countries of the 'Third World', the demographic transition is beginning. Family planning programs, organised by governments, are already having some success, especially in cities. In many such countries women recently had six children on average; now they have four. In one very populous country, Indonesia, during twenty years from 1971, fertility has fallen by 46 per cent. An important factor is the availability of several contraceptive methods. Another is the status of women: where women have begun to be freed from male domination, birth rates have declined. Among Muslim nations, Java has the best educated women and the lowest birth rate. Among Indian States, Kerala leads in the education of women and has a fertility similar to that in France and Britain.

Enquiries among the very poor of many countries have also revealed a large unsatisfied demand for family planning. Even Catholic women, despite the rulings of the Vatican, are choosing for themselves when they shall have children and how many. In

1994, at the United Nations, representatives of 170 countries made a plan, largely created by women, to stabilise the world population by the middle of the twenty-first century. It includes proposals not only for expanding birth control but also for health care and education.

What should our children and grandchildren expect? All predictions of human numbers are based on incomplete information, owing to the persistent variation in human attitudes and conduct. The findings of demographers who make the predictions often include warnings which are sometimes heeded by governments. Hence, the predictions themselves can alter the future. Here are three imagined prospects, presented in 1993 at an international conference of demographers. First, the current decline in fertility goes no further than a world average of 2.5 children per woman. In the (unlikely) absence of widespread famine or similar disaster, that would take the world population from 5.5 thousand million in 1993 to about 19 thousand million by the end of the twenty-first century.

Second, fertility declines to a global average of 2.1. This is the figure required, in the long run, to keep the population constant. Numbers would, however, continue to rise during the coming century, until they levelled off at about 11 thousand million. Third and last, fertility goes down steeply to 1.7 per woman. World population would then rise only to 7.8 thousand million; after that it would slowly decline.

The statement which contains these figures comments: 'humanity's ability to deal successfully with its social, economic and environmental problems will require the achievement of zero population growth within the lifetime of our children.'

Chapter 9

Social Lives

Beware instinct!
SHAKESPEARE

MOST ANIMALS ASSEMBLE IN social groups with others of their own kind. So do we. And the social activities of animals are often equated with those of people. But describing animals as if they had human abilities and feelings ensures that we fail to understand them. Arguments in which animals are used to explain our social lives are usually just as misleading. Yet we can learn from such comparisons.

Animals Are Not Human

Even describing what animals do is difficult, because we use words originally applied to people. (Sometimes, this is unavoidable.) The largest individual in a colony of termites ('white ants', Isoptera) may be twelve centimetres long: this monstrous creature, the only fertile female in her colony of millions of tiny creatures, lays an egg every two or three seconds. She is called the queen. The single fertile male, the size of a wasp, is called the king.

Nobody confuses a queen termite with a human queen. In other contexts, however, it is easy to make mistakes. To us a cage looks like a prison while, for an animal, it may be a home and a refuge. A Swiss zoo director, H. Hediger, has described how a herd of roe deer (*Capreolus capreolus*) 'escaped' from a paddock in the Basel zoo into a nearby forest. The gates were left open and the next morning the whole herd had come back. In experiments, hens from battery cages have been given a choice between their cages and an open space; some chose the cages. Many animals in nature occupy only a narrowly defined region and may resist being removed. Try driving a European robin from its territory. Animals have no concept of freedom.

Nor do they possess human intelligence. Yet some have been credited with arithmetical ability. Among them were the Elberfeld horses, trained in Germany and Austria early in the twentieth century. The most celebrated, Clever Hans, belonged to a teacher of mathematics. Interrogated in German, the horse answered numerical and other simple questions by stamping a hoof or pointing with his head. Thorough investigation ruled out fraud, for Clever Hans could perform in the absence of his trainer. Experiments showed him to be responding to cues, usually slight movements of the head, given unintentionally by observers who often wore hats with broad brims. If nobody present knew the answers, the horse was baffled.

The social interactions of animals can be still more confusing. In a popular story, a subordinate male wolf (*Canis lupus*) is described as presenting his neck to a dominant male, as if he were a human being making a symbolic gesture of submission. This patently absurd notion too was exposed by careful observation. The wolf presenting his neck or flank to another also growls and bares his teeth: he is the *dominant* male; the subordinate cowers or turns away. The original story was an invention. The real behaviour is typical of dominance within a group.

To understand animals, therefore, we first need exact descriptions. This objective method is owed above all to Nikolaas Tinbergen (1907–1988), a Netherlander who, as a student, was notable mainly as a hockey player. When he had taken his degree, he rejected the advice of his supervisor and took up the then unfashionable field of behaviour. Among his

'Threat' and 'seduction'. Signalling by a male South American spider, *Corythalia xanthopa*. Above, the posture adopted toward an intruder. Below, attitudes taken during courtship.

Animal signals

'Look out!': alarm signals are common among birds, mammals and others.
'Go away!': such 'threats' occur in territorial and dominance interactions and when a female weans her young.
'I love you!': prolonged and elaborate courtship is widespread.
'I'm friendly!': an animal may signal and so avoid a clash.
'Help!': distress calls are heard when a young bird or mammal is in need of food or warmth.
'Come here!': a parent may signal to the young.
'This way to food!': ants, bees and others guide members of their colony, especially by odour trails.

researches were ingenious experiments on the signals used by animals in their social interactions.

Animal signals often include complicated sounds or movements. Yet most of the messages they convey are simple. They correspond to what, in human affairs, we should call exclamations of feeling or desire or demand. Above is a list of such messages, translated into English.

SUPERSOCIALITY: THE BEES AND THEIR DANCES

The honey bee (*Apis mellifera*) provides a special instance of the last item. Termites, ants and bees are the three kinds of insects we call eusocial (supersocial would be better): they cooperate in caring for young; they have division of labour (queens and drones accompanied by sterile workers); and at least two generations live together in the same colony. They also seem incessantly active. 'Go to the ant, thou sluggard; consider her ways and be wise.' That exhortation, from *Proverbs*, is an example of the traditional use of animals to back up moral teachings.

Honey bees, like termites, have a prominent queen. (She was called the king until, eventually, it was pointed out that she lays eggs.) If she is removed from the hive, the behaviour of the others changes. An observer in the eighteenth century said that he could detect, in the altered buzz, 'a low, mournful lament'. If, however, a queen bee is isolated in a miniature cage, inside the hive, and the workers can touch her, no disturbance occurs. (It is as if the leading personage of a human community were confined in the marketplace; but the citizens carry on as usual, so long as they can occasionally shake her hand.) The queen bee produces secretions (pheromones) which regulate the workers' behaviour.

The barely credible signals of bees were first fully revealed by Karl von Frisch (1886–1982), a quiet Austrian who, when young, became embroiled in controversy. He had studied the colour vision of fish, such as minnows (*Phoxinus*), common in European streams. A very dogmatic German authority, Karl von Hess, however, stated that all fish and all invertebrates are colour blind. Experiments by von Frisch clearly showed the contrary; but von Hess refused to be convinced and instead slandered his junior. The truth was generally accepted only after years of debate.

In the past, the 'busy' worker bee has been used, like the ant, to urge us to be industrious and cooperative. Today she is known to spend much time wandering around in the hive. The adult starts off as a house bee: she explores, cleans empty cells and builds or repairs cells with secretions of one set of glands; she feeds larvae with secretions from another set; she takes and stores nectar or pollen from foragers; she may also, for a time, become a guard bee and drive away strange bees.

After about three weeks she becomes a forager. Orientation flights enable her to find the way to sources of pollen or nectar and then return to the hive on a bee line. She may also bring in water. Bees can be trained to fly to a disc coloured, say, blue

The waggle dance of the honey bee indicates the distance and direction of food to other worker bees which cluster round. Communication is aided by sounds and odours. A dancing bee conveys a message of the following form: 'I have found a new source of food, which tastes like this, 300 m south-west of the hive.'

and to ignore a nearby yellow disc. The reward for a correct choice is sugared water, which corresponds to nectar from a flower. Bees learn quickly and remember quite well for at least two weeks.

To see what bees do in private, von Frisch designed a hive with glass walls. When a foraging bee returns with food (or water), she sometimes runs round in circles in the hive (the round dance); at intervals she also transfers some of the loot to other bees. The round dance is usually performed when a new source of supply is not more than 40 metres away. Other bees fly out and collect food from the same flowers. The really startling behaviour is seen when a new source is more distant, perhaps several hundred metres. The forager then performs the waggle dance and gives off bursts of sound. The direction of the straight run indicates the direction of the food; the time spent dancing indicates its distance. Other bees assemble around the dancer and make contact. They then fly off to the place indicated. The astonishing thing is that the dancing bee has reported on something distant in time and space, and also on the nature of what she has found.

The dances of bees are, however, not like human speech: they convey only the whereabouts of certain objectives; and their form, like a map, corresponds to the meaning of the message. Bees do not use their 'language' to argue, tell stories, make promises or teach young bees. As soon as they emerge as adults, workers are equipped for social life. They do learn their way about and what flowers are sources of food. The extent to which the different kinds of activity depend on gradual learning can be discovered only by experiment.

BIRDS OF A FEATHER

A colony of bees is a single family with one fertile female, the queen: all other members are her offspring; the workers are sterile females; males (drones) are few. The family groups of birds are less strange; but the variety among them still offers plenty of temptation to those who wish to equate animal and human behaviour.

The female mallee fowl or thermometer bird (*Leipoa ocellata*) of the Australian outback lays about twelve eggs in a deep pit filled with plant matter which ferments and heats up. Both male and female regularly test the temperature by taking some of the compost in their mouths; and, for six or seven months, they keep the temperature steady at about 34 degrees Celsius by adjusting the composition and ventilation of the nest. Yet, at the end of incubation, these apparently devoted parents leave the chicks to emerge and to fend for themselves: parents and young do not meet.

Thermometer birds are not typical. Nor are the nest parasites which lay their eggs in the nests of other birds and then fly off and desert them. The European cuckoo (*Cuculus canorus*) is the most notorious but is only one of many.

Most birds form pairs, at least in the breeding season. But we still find much variation. Among the oddities are the bower birds (Ptilonorhynchidae) of forests in Australia and New Guinea. The males of the fifteen species build structures, mainly of twigs, usually in the form of a short avenue which may be 1.5 metres high. The floor of the bower is ornamented with coloured objects, such as flowers and fruit. One species prefers blue objects. The elegant bowers of some species suggest that their makers have an esthetic sense. They are, however, hardly discriminating artists: among objects they have used, or tried to use, are plastic caps, film cartons and the blue sock of a visiting naturalist. One explorer had to search for his glass eye, put aside for the night, in the bowers near his tent. The bowers are the courtship displays of the drab males: they attract females. After mating, each female builds her own small nest, lays and incubates eggs and cares for the brood.

Another oddity is the hornbill (*Buceros rhinoceros*) of southern Asia. The female spends the breeding season walled up in a tree. She incubates the eggs and is fed by the male through a hole. The male therefore looks like a ruthless chauvinist imprisoning his spouse behind a harem wall. But close observation shows the female making the wall. We are, however, not entitled on that account to call the female a domestic tyrant.

Some breeding pairs have helpers. In the Antarctic habitat of emperor penguins (*Aptenodytes forsteri*), the temperature may fall to forty degrees below zero. These birds, a metre high and up to forty kilograms in weight, do not spread themselves out in private territories. In a blizzard they huddle together and stand still. To breed, the penguins travel up to 120 kilometres from sea to rookery. The female, who has fasted for about six weeks, lays a single egg and then goes off to gorge again on sea food. The male incubates the egg on his feet, under a flap of skin; he fasts in turn and loses about 40 per cent of his body weight. In midwinter the chick hatches; the female returns and helps to feed it by regurgitating food; but an uncared-for chick is looked after by any nearby adult. Hence, by means of an improbable set of 'adaptations', emperor penguins survive apparently impossible conditions. If they are a threatened species, it is not by the climate but by human exploitation of Antarctica.

Monogamy is the rule among birds but it is not universal. The breeding male of some species has several mates (polygyny). One, *Euplectes franciscans*, also called a bishop, builds several nests and attracts a female to each. An ornithologist, who had been observing this brightly coloured species, once tried to publish a paper entitled, *Territory and Polygamy of a Bishop*, but a stuffy editor rejected the title. More surprising, some females practice polyandry. In Central America, the female northern

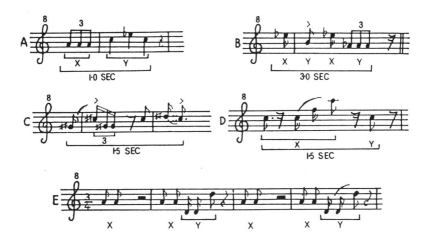

A musical aid to pairing. The duets of five pairs of African shrikes, *Laniarius aethiopicus*. Each pair has its own melody. X and Y in each song represent the contributions of the two singers.

jacana (*Jacana spinosa*) is large, colourful and noisy and might be taken for a male. She has two or more nests, each made by a different, smaller and unobtrusive mate. The males stay with the eggs while the female parades around and drives off intruders.

Mating and care of young often require individual recognition. In certain African forests a visitor may hear the musical, individually distinct and relentlessly repeated songs of the boubou shrike, *Laniarius aethiopicus*. The visitor may also, with skill and good fortune, detect that every song is a duet: each member of a pair has its own notes. If one flies off, the other may sing the whole tune; the erring member then returns. Duetting was at first supposed to keep pairs together in dense forest. Perhaps it does. But many other birds are duettists and not all live in forests. Some sing duets only when they can see each other.

In the bestiaries of the European middle ages, moralists strengthened their messages by telling tall tales about animals. As the reader can now see, modern bestiarists would have plenty of material. We should be devoted husbands like hornbills; neighbourly like penguins; dedicated to social duties like bees; artists in our homes like bower birds. Would the marriage bond be strengthened if couples copied boubou shrikes and sang private duets daily before breakfast?

Peck Orders and Status

Ethologists need not complain at literary forms devised 'to point a moral or adorn a tale'. Trouble comes when animal fables are used to *explain* human conduct. One member of an animal group is sometimes described as dominant over another. Popular writers have revelled in this phenomenon. According to one, dominance among animals explains our supposed need to follow a Leader.

Modern, authentic studies of dominance had a beginning in 1922. T. Schjelderup-Ebbe, a Norwegian psychologist who regarded dominance and its opposite, subordinacy, as a fundamental principle of animal and human existence, described the social lives of groups of domestic chickens: the most dominant individual pecks all the others; the next pecks all except the top one; and so on. Hence arose the expression 'peck order'.

A peck order is an example of a status system (or dominance hierarchy). Most Primates (the lemurs, monkeys and apes) have such a system. Hence each troop might be imagined as having a male Top Animal who bosses the others, has first choice of females and food and insists on the others getting out of his way. (In human affairs, these would represent the dominant individual's rights.) The same creature might also be expected to lead the group on the march, to drive off intruders of the same species and to protect the group from predators. (These activities would correspond to human duties.)

I doubt whether any species quite matches this picture. Even after hundreds of days of patient watching and reporting in nature, it may be difficult to decide what behaviour should be called dominant. Among the baboons (*Papio*), easily seen on African plains, harmless signals used by males, such as slapping the ground or merely blinking, often make another member of the troop withdraw. Baboon females are subordinate to males (in the sense of getting out of their way), but this is not true of all Primates. Female squirrel monkeys (*Saimiri sciureus*), approached by a male, usually drive him off; even young females 'threaten' adult males. Female ring-tailed lemurs (*Lemur catta*) are similarly dominant: a female may rush up to a male, snatch an item of food from him, and give him a cuff on his ear. Whether this is interpreted as a reproof or as an affectionate gesture is likely to depend on the prejudices of the observer.

Other kinds of relationship are listed below. One is determining the troop's direction of movement. Baboons which lead the way need not be the dominant ones. Another crucial activity is care of the young. On a long march a large male baboon may carry a baby belonging to his consort. Some adult macaques (*Macaca*) without babies hold and carry young monkeys.

Some social roles

Sexual and consort relationships
Care of young (not always by parents)
Filial conduct (not always toward parents)
Dominance and subordinacy
Leading or directing group on the move
Combining to get food
Controlling social interactions
Protection against predators
Driving away members of other groups

Closely related species may be very different. Two chimpanzee species are now recognised: the long known *Pan troglodytes* and *P. paniscus*, the bonobo (pigmy chimpanzee). Both live in African forests, yet socially they are quite apart. Unlike chimpanzees, bonobos copulate frequently, often face to face; and both females and males go in for much genital rubbing with members of the same sex: they are, in a sense, extremely gay. Among other bonobo peculiarities, females have right of way over males and prior access to food; and the female protects her son from other males. A male may remain attached to his mother's apron strings even when fully grown. Still more surprising, bonobo tool use, unlike that of chimpanzees, is not conspicuous or elaborate. Some writers have used bonobo behaviour to suggest that male dominance is not 'natural' to humanity. But, since knowledge of the conduct of one animal species does not tell us what to expect of any other, none can be expected to give us reliable information about our own social lives.

TERRITORIES AND LAND OWNERSHIP

In the list of social roles above, the last item is one aspect of territorial behaviour. Animal territories have sometimes been equated with human ownership of land. So we may ask: do animals own property? and is landholding a 'natural human instinct'?

An ideal animal territory would be held by a family throughout the year; it would contain every necessity—water, food, shelter and material for a lair or nest. It would also have places from which the holders could perform visible displays, utter sounds or waft odours—signals which drive off intruders. Of course, few territories are just like that.

Until recently, knowledge of animal territories came almost entirely from bird watching. Most of the species studied have breeding territories.

> . . . lo, the winter is past,
> The rain is over and gone;
> The flowers appear on the earth;
> The time of the singing of birds is come.

The birds not only sing: some also show off bright plumage. Both songs and visible displays are usually assumed to have two functions: attracting mates and driving off rivals. To test this requires experiments. The North American red-winged blackbird (*Ageleius phoeniceus*) sings its typical song from conspicuous posts in its breeding territory and, by spreading its wings, it shows off distinctive red epaulets. If the birds are given reserpine, a tranquillising drug, displaying is reduced and neighbours trespass more. In addition, males with their epaulets covered may lose their territories altogether. Another kind of experiment has been used on a European bird, the great tit (*Parus major*). If territory holders are removed, they are quickly replaced by birds from a floating population of 'landless' individuals. This invasion can be delayed by installing loudspeakers which broadcast the songs of the absent birds.

When humanity is described as a territorial species, the argument (if there is one) is this: many animals are territorial; human beings are animals; therefore, we are territorial. But not all animal species hold territories. A popular writer or a moralist who wishes to *deny* that territorial conduct is natural for people may fall back on rhinos; or on monkeys such as the red colobus (*Colobus badius*); or on gorillas.

Popularisers regularly use this method. Each territorial species has only one kind of territory. They choose species whose habits match their message and ignore others with quite different habits. They also disregard the variety of human societies. The human species, during most of its existence, has consisted of small bands of gatherer hunters with no property as we know it. Only since the development of agriculture have most human societies been based on ownership. Some began with communal occupation of land. In many regions this gave place to feudal order in which a few families controlled large estates while most people had none. Some of the longest lasting and most prosperous civilisations have depended also on owning persons, that is, on slavery. Today slavery is illegal and abhorred. We do, however, retain private ownership of several kinds and also public tenure. These are regulated by changeable laws and customs. This cultural variety has no counterpart in any one animal species.

When animal territories are equated with human property, we are told, in effect, that owning property is a fixed, instinctive feature of our social lives and that attempts to change it are against nature. Some economists and social scientists argue that, for biological reasons, the kind of ownership we accept today is inevitable. Such arguments are not biology or social science: they tell us only what kind of society is preferred by the writer. They have no relevance to the merits or defects of the writers' social proposals and they do not explain human social life.

CLASHING CROWDS?

Comparisons can, however, be helpful. Ideas about human crowding have come from findings on animal populations.

In a favourable region, the members of a species spread themselves out. Such behaviour ought to keep numbers fairly steady, yet some territorial species fluctuate widely. The Norwegian lemming (*Lemmus lemmus*) is reputed (wrongly) to form suicidal crowds every few years and to swim out to sea and drown. Lemmings make one of a number of species described as 'cycling'. Many species of voles (*Microtus*) of Britain and Europe and the deermice (*Peromyscus*) of North America hold individual territories in grasslands, where they may be difficult to find. But, every four years or so, they become numerous;

they may then swarm away from their usual areas into pastures and forests. But a 'crash' follows and, in a few months, most disappear. This can occur when food is plentiful, predators are few and when no unusual infections are found. So, for a regulating factor, we turn to behaviour. Apparently, interactions which cause the animals to spread out increase with population density. Wounding may become frequent, the fertility of females may decline and death rates among the young go up.

Social scientists have therefore suggested human crowding as a cause of violent crime and other ills. The densest crowding, however, usually goes with many other conditions such as poverty, bad housing, unemployment and poor education. The best researches have found little or no ill effects of crowding itself. The quality of housing is more important. Nor does crowding reduce human fertility. People often like to be crammed together (as in holiday resorts, where crowding does not lower the birth rate); and, as we know, isolation is bad for most human beings. In this, people may seem to resemble the mammals of closely packed herds. But social scientists rarely base their researches on studies of sheep, bison or zebras.

Although the original hypothesis, that of the bad effects of crowding, has not been confirmed, the method used was valid. Studies of animals suggested what could be looked for in human communities. Many useful findings resulted. One is that crowding can, in some circumstances, *reduce* violent crime: the reader is at greater risk in a quiet street than in one with plenty of people. More important, violence does rise with unemployment, poverty and lack of education.

'INSTINCT'

Much of the animal behaviour described above is commonly said to be instinctive; and in Shakespeare's *Henry IV, Part 1*, Falstaff warns us to beware of instinct. He was quite right: the idea of instinct regularly causes trouble.

Recently, it was usual to divide behaviour into two distinct kinds: some writers contrasted instinctive with intelligent acts; others used the words, innate and learned. The animal behaviour which used to be called innate seems to have an automatic character, as if it were switched on by a particular situation.

When a cuckoo chick (*Cuculus canorus*) hatches in the nest of a meadow pipit (*Anthus pratensis*), the cuckoo makes violent movements and throws out the pipit chicks. The parent birds then feed the cuckoo, even if their own young are squeaking loudly outside the nest. The large gaping beak of the cuckoo chick is all the stimulus they need. (A famous English zoologist, Julian Huxley (1887–1975), once suggested a rather disagreeable human parallel: a mother devotedly attends to a young gorilla, while her baby is screaming outside.)

Such behaviour has been contrasted with learned or intelligent action. This sharp distinction, however, often breaks down. Some animal behaviour seems to be unalterable but is not. Many birds and mammals can walk or run soon after hatching or birth. They then follow their mothers. This apparently fixed performance depends in fact on learning to recognise the mother. The learning can go wrong. The reader will recall Mary's little lamb: 'everywhere that Mary went, that lamb was sure to go'. The lamb ought to have been sheepishly or 'instinctively' following its mother.

Ducklings (*Anas*), like lambs, follow their mothers. In experiments, some have been hatched in refrigerators. Then, instead of meeting a duck, they have been allowed to see a plain box. They followed the box as though it were a parent bird. (They followed more readily if the box squawked like a duck.) Learning to follow a parent, or an object as if it were a parent, is imprinting. It is not a wholly passive process. A newly hatched duckling left alone utters distress calls and moves about. But, if it is shown an object, it runs toward it and stops calling. It then gradually learns to recognise and to follow the object from any angle, whether it is near or far. Ducklings become imprintable about eight hours after hatching and remain so for about another twenty hours.

Sexual behaviour too may depend on early learning. Young drakes have been brought up with birds of a different species. When adult, they try to mate with ducks of the species with which they have been reared. The sensitive period for developing such abnormal behaviour is four to nine weeks from hatching. Both imprinting and sexual imprinting have been found, by elaborate experiments, among mammals as well as birds.

Another such case is bird song. The complete song of a common European bird, the chaffinch (*Fringilla coelebs*), develops during the first thirteen months or so from hatching. A young chaffinch sings the normal song only if it has the opportunity to imitate an adult during that period. Chaffinches have been reared in soundproof rooms during their first year. They then never learned to sing normally. Birds can also be reared in small groups, isolated from other birds. Each member of the group then sings a melody similar to that of the others and different from the usual song. Many other species of birds learn their songs early in life.

Both imprinting and the development of some bird songs illustrate a fundamental rule. Without experiments, it is usually impossible to say how any kind of behaviour develops. The behaviour called instinctive or innate may appear gradually as the animal grows. If an animal develops in an unusual environment, its apparently fixed behaviour may be abnormal too. Normal, species-typical conduct may depend on quite a long process of learning. Once again, we see how important it is not to assume that a trait is fixed by the genes at conception.

HUMAN SIGNALS AND SOCIAL LEARNING

Expressions such as instinct and innate behaviour are therefore going out of use, especially in accounts of human social action. Can we nonetheless still find traces of what Darwin called 'the indelible stamp of our lowly origin'? In fact, a few items of our nonverbal communication, especially those of infancy, develop reliably in most environments. A hungry infant utters a distinctive cry, just as other mammals do; an infant in pain has a different cry. A more welcome signal of infancy, 'social smiling', typically appears at perhaps four weeks and continues, for some months, as a response to any human face. An infant may even smile at a hideous caricature of a face. (The reader can try the harmless experiment of waving one in front of a child of, say, four months.) Later, many infants smile only at familiar faces and turn away from strangers. Although every infant has a distinct personality, smiling and crying are universal. They are typical of *Homo sapiens* and can help pediatricians to judge how well an infant is developing.

Adults too have a few universal signals, such as laughter, but most of our nonverbal communication is learned from the people around us. Probably, the reader indicates 'yes' by nodding the head up and down; but this means 'no' in some communities. Kissing too varies greatly: footballers and even cricketers now embrace each other in public when a goal has been scored or a wicket taken. In some communities where it is now usual, such conduct would recently have been quite shocking.

All complex human social action is founded on social learning, especially imitation and teaching by others. Here is a list of the main kinds of social learning.

Kinds of social learning

'Imitation'
Stimulus enhancement (drawing attention to something)
Emulation ('making a splash')
Imitation in the strict sense

'Teaching'
Encouragement (offering an opportunity)
Deterrence (keeping from danger; also driving away)
Teaching in the strict sense

Several kinds of social learning are familiar from the play of children. A small child sees another playing with a toy; the child then plays with it also because attention has been drawn to the toy. This is called stimulus enhancement. Or one child may see another achieve something, such as making a splash by throwing an object into water; the first child then does the same thing. This may be called emulation ('I can do it too!').

Casually, we might say that both are examples of imitation; but the word imitation is better reserved for something more complex: an individual has to observe another perform an unusual act and then go through the same novel sequence. A human example would be a joiner shaping two pieces of wood, drilling holes in the wood and using screws and screwdriver to fasten them together: the apprentice watches and does the same with more wood.

Much human learning depends not only on imitation but also on teaching by others: the skilled person *persists in instruction until the pupil achieves a certain standard of performance or improvement.* Social transmission of skills and habits leads to the development of local traditions. These are passed on, from generation to generation, not by genes but by example. In human societies, tradition is an essential element of our social conduct and our skills. Moreover, transmission is not only from elders to young: it is often among individuals or groups of the same generation.

To what extent do animals imitate each other; and, still more interesting, do they teach their young? The most likely candidates for doing so are the apes. This obviously applies to the manufacture and use of tools. Hence much attention has been given to animal species, not only apes, described as tool users. Among them are wasps which tamp down earth over their nests with a pebble. A bird (one of Darwin's finches, *Cactospiza pallida*), uses a twig to get at its insect prey from under bark. The sea otter (*Enhydra lutris*), in the North American Pacific, swims on its back and bashes mollusc shells on a rock balanced on its belly. Only one kind of tool, however, is used by each species and these animals are not, as we are, tool *makers*. Nor is the behaviour a product of social learning: it resembles building a characteristic nest or digging a typical burrow.

Only chimpanzees (*Pan troglodytes*) achieve more. They use sticks or stones for getting food. They also throw them at other chimpanzees, at baboons and at people. They use bunches of leaves as sponges for getting water and for removing blood from wounds or for wiping their bottoms; and they pick their teeth with twigs. They have been described as having tool kits. Moreover, separate populations of chimpanzees differ in the ways in which they capture insects and in their use of tools. Some eat ants but others eat termites. Both require the use of carefully chosen and adapted sticks but in each region the details are different. Some groups eat nuts but again methods vary: nuts may be cracked with a club or a stone, on an 'anvil' which may be a root or a rock.

So chimpanzees are not only more versatile and constructive than are other tool users: they also have local traditions and have therefore been thought to imitate as human beings do. The young undoubtedly have their attention drawn to possible tools; but each, it seems, has to learn skills for him or herself. For this, a strong tendency to investigate objects is important: chimpanzees are inquisitive. The most careful studies have revealed no clear case of imitation in the full sense. Nor do chimpanzees systematically instruct other chimpanzees in the performance of skilled acts. Despite the assertions of some writers, we should, as usual, apply common sense and reject the notion that a teacher in a classroom or lecture theatre is faced with a troop of grimacing simians.

In teaching and social learning, therefore, a wide gulf exists between simian and savant and even between ape and student. As we see later, in chapter 11, this gulf becomes still more evident when other human peculiarities are examined.

Giraffe (*Giraffa*). What is the function of the long neck? And of the distinctive markings? Kipling's *Just So Stories* are not a reliable guide.

A Struggle for Existence?

How did the millions of species, past and present, arise and survive? The only convincing answer we have is: by natural selection. But this is not a simple idea; nor is it a complete answer. Commonly used expressions such as fitness and adaptation can be deceptive; and, as in the previous chapter, violence—often imaginary—insistently raises its head. So do questions on the implications of evolutionary theory for our understanding of ourselves. Some arise from slogans such as 'the survival of the fittest' and 'the selfish gene'.

CHAPTER 10

THE NEW DARWINISM

> When we reflect on this struggle, we may console ourselves with the full belief that the war of nature is not incessant, that no fear is felt, that death is generally prompt, and that the vigorous, the healthy and the happy survive and multiply.
>
> CHARLES DARWIN

GIVEN THAT EXISTING ORGANISMS have arisen from very different forms in the past, how did they do it? Despite Darwin's cheerful comment above, should the heading of this chapter be Tennyson's famous phrase: 'Nature, red in tooth and claw'? Is evolution a product of war in the natural world? A widespread belief exists that even human beings, as a result of their evolution, are inherently violent.

Before modern biology began, Christian writers had been inspired not by violence but by the seeming perfection of nature. Previous chapters give many examples of how organisms match their environments. The most conspicuous traits are structural: the mechanical perfection of a bird in flight or of a fish swimming, or the exact fit of flower and pollinating insect.

Other such features are behavioural. In 1762, H.S. Reimarus (1694–1768), a German professor of Hebrew and philosophy, published an encyclopedic account of the instincts of animals: by which he meant their many activities, apparently skilled, but not acquired by imitation or other kinds of learning. Nest building, courtship, parental care, choice of food, avoidance of enemies and much else, Reimarus regarded as direct proofs of the existence of God: how could animals know what to do without divine help? The same theme was powerfully taken up by an English bishop, William Paley (1743–1805). He held both instinctive behaviour and also complex structures, such as the eye, to be evidence of a Creator. Paley's concept was of a divine designer, operating in a way recognisably human yet with superhuman skill. It recalls the ancient Greek saying, that man makes gods in his own image. When we refer to the *design* of organisms we are resorting to metaphor: like the reductionist writers encountered in chapter 7, we are speaking as though living beings were mechanisms created by the planned exertions of an engineer.

Lamarck and the 'Inheritance of Acquired Characters'

When the idea of organic evolution began to replace that of a Creation, the fit of organisms to their environments was at first attributed to the inheritance of acquired characters. Biologists, however, have for decades regarded this doctrine as a rather disreputable hypothesis to be briefly dismissed at the beginning of any account of evolutionary theory. Yet from time to time somebody comes forward with questionable evidence in its favour. Hence it has to be regularly shot down.

We associate the idea especially with the French biologist, J.B.P.A. de Monet (1744–1829), generally known by a title, Chevalier de Lamarck. Monet was destined for the priesthood but, on the death of his father, he joined the French army. He was immediately, at the age of seventeen, involved in a heroic action and promoted. When the Seven Years War ended, he resigned his commission. For fifteen years he lived in Paris as an impoverished writer. He also became an enthusiastic botanist, wrote a successful work on the flora of France and eventually, at the age of fifty, was appointed a professor of zoology.

Lamarck was unusual for his time in being both irreligious and an evolutionist. Evolutionary change, he thought, was brought about by the effects of use and disuse of organs and by the exertions of organisms in adapting themselves to environmental demands. According to him, and to many other biologists of his time and later, the results of such influences

Lamarck

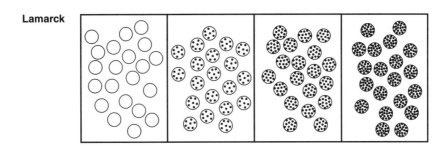

Darwin

mutant individual

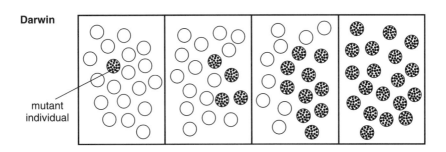

'Lamarckism' versus 'Darwinism': how does a population of bacteria develop resistance to penicillin? Lamarckian individuals (above) adapt to a changed environment and transmit the result to their offspring; the next generation does the same, only more so; and so on. The Darwinian account invokes natural selection. Some variation always exists in a population, and some is genetically determined. Below, a mutant is present and, on exposure of the population to penicillin, this unusual form survives longer than the others and has more descendants. Eventually, all that remain are descended from the previously rare form. The effects of individual adaptation to change (if any) are not passed on.

are transmitted to the individual's offspring. Even Charles Darwin believed this.

Modern knowledge makes Lamarckism seem extremely improbable. Suppose an animal develops thick calluses on its toes and extra strong muscles, through much running on hard ground. According to Lamarck, these features should reappear, at least partly, in the animal's young *before* they begin running. For this to happen, the animal's genes would have to be so altered as to promote the early growth of calluses and muscles. Similarly, when a person learns to be a good pianist, the person's genes should change correspondingly: children subsequently conceived could then become pianists with less training. Another difficulty is that not all acquired features are advantageous: during an individual's life, injury and disease may cause impaired function. It would not be adaptive for such adverse changes to be passed on.

Even the exactness of 'adaptations' tells against Lamarckism. An evolutionist, John Maynard Smith, gives the example of an orchid which looks like a bee and so attracts male bees; the bees get food for themselves and also ensure pollination of the orchid. The flowers of one species, seen from the side, have the shape of a bee and even a red mark corresponding to a bee's eye; they also smell like a female bee. Such features could hardly result from the direct action of the environment on ancestral orchids. Nor can we suppose that orchids ever *strove* to achieve such an appearance.

NATURAL SELECTION: STABILITY AND CHANGE

The alternative to Lamarckism is natural selection. We owe the development of this idea especially to Charles Darwin (1809–1882), but it was thought of independently, at about the same time, by A.R. Wallace (1823–1913). Both were English. Darwin was the son of a physician. He attended the universities of Edinburgh (where he began to study medicine) and of Cambridge (with the intention of entering the church); but he was an unsatisfactory student in both. His father's scorn for his playboy son is notorious. 'You care for nothing but shooting, dogs, and ratcatching, and you will be a disgrace to yourself and all your family.'

At the age of 22 Darwin overcame paternal opposition and embarked on his famous voyage of five years in *Beagle*. He is usually described as the ship's unpaid naturalist but he was, it seems, also taken on as a companion for the captain, Robert Fitzroy—somebody to talk to on the long voyage. It was fortunate that Darwin had not yet formed his ideas about evolution, for Fitzroy was a fundamentalist Christian and later a strong opponent of Darwinism. The captain's tasks were to chart the coasts of South America and to attempt accurate measurements of longitude. For the latter he was equipped with 22 exceptionally accurate chronometers.

The voyage allowed Darwin to study the natural history, geography and geology of a large part of the planet. His life on his return continued to be quite unlike that of a modern scientist: he had no formal qualifications as a biologist, no university post and no laboratory; above all, he never had to worry about money. An additional contrast was the scope of his researches. In *Beagle* he was as much a geologist as a naturalist. His most famous contributions were theoretical; but he also made a complete study of a group of crustaceans, the Cirripedia (barnacles and others), and he published important reviews of the domestication of plants and animals and of human evolution. In two fields of research, on earthworms and on climbing plants, he carried out or designed ingenious experiments.

Darwin's most important publication, *On the Origin of Species by Means of Natural Selection, or the Preservation of Favoured Races in the Struggle for Life*, first appeared in 1859. There he presents his theory of natural selection and also evidence that evolution has actually occurred. He had been labouring for nearly twenty years on a longer work when, in 1858, he was shocked at receiving a paper which proposed a theory identical in substance to his own. The author was a distinguished naturalist, Alfred Russel Wallace, who was working in the Malay Archipelago. Both men displayed generosity. In that year a joint paper was presented to, and published by, the Linnean Society in London. Later, it was Wallace who insisted that the theory should be called Darwinism.

Today, especially in the form of slogans such as 'the survival of the fittest', natural selection is a familiar notion. But, in 1859, supporting the idea of evolutionary change and explaining it by the impersonal, amoral process of natural selection was bold, almost foolhardy. In other societies, Darwin and Wallace might have been burnt at the stake or stoned to death as heretics.

Even among those who accepted evolution, Darwinism has been a source of difficulties and controversy. Darwin's successors have taken decades to work out, even partly, where it leads. Major advances in theory began to be made more than sixty years after the *Origin* was published. Among the founders of modern Darwinism, two of the most famous make such a contrast, that they sound like characters in a work of fiction. One, R.A. Fisher (1890–1962), a small, quiet, bearded Englishman, was an original mathematician. He was also interested in human genetics, both scientifically and as a social phenomenon: at the beginning of the century he helped to found the Eugenics Society in London. Members of the society were worried because the poor were breeding faster than the people who, they believed, had much better genes. In 1910, Fisher stated that the upper classes, because of their heredity, contain 'all the finest examples of ability, beauty and taste': he urged, in effect, the survival of the richest.

The other, J.B.S. Haldane (1892–1964), one of a prominent Scottish family, was a large, formidable man who rowed for his college at Oxford University, was an infantry captain in the first world war and had a small, military mustache. He was also a polymath of disconcerting brilliance: he was a classical scholar, a biochemist who experimented on himself and made a substantial contribution to the study of enzymes, a geneticist and, like Fisher, a mathematician. He inspired large numbers of young people, including myself, with his many popular essays. He was also energetic in leftwing politics. And he was a devastating critic of the biology and politics of the eugenists: the eugenics movement, he once said, was a passionate protest against the hard fact that the meek do inherit the earth.

Both men made original contributions to the understanding of natural selection. To explain the new Darwinism, we begin with three facts. First, all species can reproduce at a rate much higher than is needed to keep the population steady. The unchecked reproduction of even the slowest breeders, such as elephants and human beings, could—after some years—be overwhelming. But, as we know, high rates of breeding are eventually countered by high death rates or failure to breed. Second, individuals in each population vary; some of the differences are genetically determined; and some genetical types survive longer, and breed more, than others. Third, mutation ensures that genetical variation is always present. But we now face a difficulty. Departures from the typical or normal (for instance, white colour in a usually dark species) are usually disadvantageous. How then can they be a source of progressive change? Natural selection, it seems, *preserves* existing forms. Such stabilising selection is often found.

It was the achievement of Darwin and Wallace to see natural selection as a source also of progressive, adaptive modification. Today we have direct evidence of short-term change in free populations. The most famous example concerns the peppered moth (*Biston betularia*), which usually has a speckled appearance. The lichens on the trees where the moths settle are light

Natural selection in action: the famous peppered moth (*Biston betularia*). On the left, the usual form, difficult to see on trees covered with lichens. On the right, a melanic from 1915. More recent melanics are even darker and are hardly visible on tree trunks denuded of lichens.

coloured but speckled. The moths are therefore well camouflaged. During the nineteenth century, collectors sometimes found dark (melanic) specimens which showed up conspicuously: they make a good example of a disadvantageous abnormality, for they are quickly picked up by birds; selection in this case *prevents* change.

But, during the past century, in many industrial regions of Europe and Britain, the lichens on trees have been killed by poisons in the air. Against the bare, dark trunks, the speckled moths now make sitting targets for predators. In such areas, the black form of the moth became the common type: only about five per cent of moths were still speckled. Direct observation confirmed that speckled moths are less eaten by birds wherever lichens remain; melanic individuals thrive only in the absence of lichens.

Breeding in the laboratory showed the black colour to be due to a simple genetical difference: melanism is a dominant trait. The story so far is therefore straightforward. Pollution produced a new environment. Among the peppered moths a very few were black, owing to the presence of a mutant gene. In the altered conditions, to be black was advantageous; and, after a few generations, the mutant gene largely replaced the alternative. As might be expected, thorough research has revealed complications. Melanics collected in the nineteenth century have small white patches, whereas the more recent ones are almost completely black. Probably, selection for the original gene has been followed by selection for modifier genes which make the blackness more uniform.

Industrial melanism is not a peculiarity of one species.

Populations of other moths and of bugs (Hemiptera) have gone the same way. They illustrate the genetical variation always present in each species and show how it can produce an adaptive change when the environment alters. (This does not imply that a population can change in any direction. The orchids which look and smell like bees do not buzz and seem unlikely ever to do so.)

The appearance of new varieties, adapted to an altered environment, can be of economic or medical importance. DDT began to be used against insects during the second world war. It made possible elimination of body lice and so of louse-born typhus—an unprecedented achievement of immense value for armies and for public health generally. Later it had an impressive success against house flies (which, too, are disease carriers). But, after a time, the surviving flies proved to be resistant: to kill them required dosages up to a hundred times those effective at first. Resistant mosquitos and lice also appeared. (Compare the story of malaria in chapter 5.) In environments dripping with insecticides, formerly rare types had evidently become common.

The use of antibiotics has met the same difficulty. Penicillin, the first antibiotic, at first miraculously effective against bacterial diseases which had often been fatal, soon met resistant strains of bacteria. Today we have an array of antibiotics with different properties which, to some extent, counter the ability of disease organisms to vary. But it is increasingly important not to use antibiotics in ways that encourage the appearance of resistant strains.

Such changes, well within a human lifetime, show one thing clearly: potentially useful variation is always available.

JUMPING GENES BREAK THE RULES AND UPSET THE PUNDITS

In biology, nothing can be taken for granted, not even the fundamentals of biological theory. In mid century, an example was provided by an American geneticist, Barbara McClintock (1902–1992). She was a sharp-tongued, dedicated experimentalist indifferent to petty rules and also handicapped by being a woman. She nonetheless commanded admiration and affection among colleagues who knew her well. By 1939, though without a secure post in either teaching or research, she was recognised as the leading authority on the genetics and chromosomes of maize (corn, *Zea mais*). She also had a marvellously close acquaintance with minute variations in the appearance of individual plants.

Her powers were most dramatically shown when, in the 1940s, she observed anomalies in the colours of her plants: seedlings which should have been green had white or yellow regions. The cause was, it seemed, gene mutations which took place during development. Morover, close study suggested that some particles, which resembled genes, could move from one position to another on a chromosome. This apparently wild suggestion seemed to contradict a basic principle of genetics—that the genes in the DNA molecule are in a fixed order. To achieve recognition of her theory, Barbara McClintock therefore needed much tenacity during the quarter century from 1950, for her findings were at first brushed aside. But today the existence of 'jumping genes' is not in doubt and her work is correspondingly honoured.

Some apparently inert (noncoding) regions of the DNA molecule are now known to be sites of mobile DNA sequences, called transposable elements or transposons. They resemble viruses, are diverse and numerous and have been found in a great variety of species, including bacteria, yeasts, insects and mammals. They are, it seems, a source of unexpected instability in the DNA of, perhaps, all organisms. What induces their movements is not known but, in plants, adverse conditions such as cold seem to increase their rate of transfer.

Movement of a transposable element can make an ordinary gene mutate. In *Drosophila* chromosomes, the arrival or departure of a noncoding chunk of DNA (an insertion sequence) is believed to be responsible for many apparently spontaneous gene mutations. Evidence has also been found of such effects in human beings: they evidently cause mutations which lead to neurofibromatosis (an abnormality of nerve tissue) and hemophilia.

Like some viruses, transposable elements can also move out of cells without destroying them. Evidence has now been found of movement between different organisms and even between different species. Fruit flies are sometimes parasitised by mites. These minute creatures seem to have transferred transposons between two species of *Drosophila*. Something like gene engineering is therefore perhaps already going on in nature. If this is confirmed, a large new region of genetics will be opened for exploration.

In addition, the question of the inheritance of acquired characters has been revived in a new form. Controversy about Lamarck's original ideas is no longer an issue for biologists: the conclusions given at the beginning of the chapter have not been contradicted; nor has the concept of natural selection been rejected. The discovery of jumping genes has, however, suggested a novel way in which environmentally induced changes can be transmitted. A cell entered by transposable elements in the way proposed is not only altered: it also transmits the changed state to its descendants. Probably, many such infections have occurred and are still occurring. Such passages between organisms can alter genes in the eggs or sperms of an infected organism and so produce a phenotypic change which reappears in later generations.

Of course, this is not like anything Lamarck could have imagined: it is a novel kind of effect which does not belong in any conventional biological category. It is, in fact, a new source of transmissible variation. Its full scope has still to be discovered.

ADAPTATION AND FITNESS

The version of Darwinism presented so far may seem disappointingly lacking in drama. I have said nothing of the war of nature, the struggle for existence or even of competition; the survival of the fittest has been dismissed as a mere slogan. Nor, to quote J.B.S. Haldane, has 'Nature' been represented 'as a stern but beneficent female who "weeds out" the unfit in each generation'. Yet these and similar expressions have been common in writings on evolutionary theory. They transfer our social (or antisocial) activities, such as competition and war, to nature; or they suggest that our agricultural practices, such as weeding, are to be found throughout the natural world. Some writers seemingly convict 'Nature' of moral deficiency, as in the phrase, 'the selfish gene'.

Why are such metaphors not used in this book? Think of two plants, each growing from seed in an arid zone. One has a greater capacity to resist drought than the other and the difference is genetically determined. Both give rise to offspring which tend to resemble their parents; but one grows larger, lives longer and produces more seeds. Eventually, after generations, only descendants of that plant remain. This is not because the two original plants were selfish, came to blows or threatened each other; nor is it because some supernatural agency acted as a gardener and picked over the growing shoots. As among the melanic moths, the change in the population results from the different chances of survival of two types. In another environment, with more rain, the advantage might have been reversed. Evolution by natural selection depends on such chances. 'Nature' can evolve without being 'red in tooth and claw'.

In the example, one plant is fitter than the other; or some would say that it has superior Darwinian fitness. The plant population undergoes a process of adaptation—in the first case, adaptation to drought. These two words, fitness and adaptation, commonly used in biology, can cause confusion. Fitness in biology is not a single quantity: it is not like the time a person takes to run a hundred metres or the weight a person can lift. In biology, the fitness of an organism is a measure of its chances of leaving descendants. Suppose we are investigating buttercups or bears: when we say that one individual or type is fitter than another, we may be referring to the number of seeds which survive and grow; to length of life; to the number of young born; or to any one of many other possible measures. Which measure is chosen depends on the species studied and what we are trying to find out. Hence the phrase, 'the survival of the fittest', is not very informative: taken literally, it means 'the survival of those that survive'; taken a little more seriously, it states the obvious: some types are more successful than others.

The more successful forms may be described as being better adapted than the alternatives. In biology, adaptation has two distinct meanings. First, it may refer to changes in one individual in response to its immediate needs. The shape of a tree and the growth of tissues in its trunk can be affected by prevailing winds; the tree is then less likely to be blown over. When a human being moves from a home at sea level to one at 3000 metres, blood-forming tissues become more active: the number of red cells in each litre of blood rises and the person can perform vigorous exercise, despite the lower partial pressure of oxygen in the mountain air. These are developmental changes which involve the growth of tissues. Among the many other kinds of individual adaptation, animals adjust their behaviour to their needs and exposure to a disease organism may result in the development of resistance.

Second, 'adaptation' may refer to a genetical change in a population which makes survival more likely. This is its usual meaning in accounts of natural selection. The melanic moths represent the simplest case: a single feature, previously rare, becomes common.

When, however, a population responds genetically to a new environment, changes are unlikely to be only in a single trait, such as colour. We can see this in the laboratory. I have kept house mice for many generations in refrigerators. The mice were 'selected' by being left to breed in the cold (near zero degrees Celsius). At first, in these conditions, many females were barren. Those that were fertile produced fewer young than did control mice kept in a warm environment. But, after only ten generations, the 'Eskimo' mice in the cold were distinctly different from those kept at room temperature: they were heavier, produced larger litters, secreted more concentrated milk and looked after their young better. These findings match what we know about the geography of *Mus domesticus*. It is present in almost every environment where people live, from the tropics to subantarctic islands. It has many varieties. In some tropical islands adults reach only about 16 grams, but in one very cold place they achieve 40 grams. If such diverse populations were isolated for a long time, they would perhaps evolve into separate species.

Large scale, progressive modification can, however, be followed only in the fossil record. The mechanical marvel of a bird in flight is the culmination of a long series of changes: feathers replace scales; the shape of the body alters; the bones become lighter; the tail disappears; teeth are replaced by a beak; and much else. Similarly, over millions of years, horses became larger and also better at running and chewing. Adaptive change usually involves alterations in many genes and in many features.

ARE GENES SELFISH?

A modern counterpart of the nineteenth century catchphrase, 'the survival of the fittest', is 'the selfish gene'. In a popular book with that title, Richard Dawkins (already encountered in chapter 7) applies the concept of natural selection to animal behaviour and, like some other writers, enlivens his text by using words we usually apply to people, especially those we dislike or deplore: for example, cheat and sucker.

Evolutionary change due to natural selection is an *automatic consequence* of genetical differences in fitness. How then can the idea of selfishness (or other moral expressions) be applied to such a process? Survival requires that an individual's characteristics benefit that individual. Organisms must not be expected to help others at their own expense; their social interactions must (it seems) all be of a kind which, in ourselves, we call selfish. To use 'selfish' in this context transfers a moral judgment to an impersonal process in nature: it is a metaphor, like the expression 'the cruel sea'.

The only obvious exception is parental care. Natural selection favours the care of young when such care makes the chances of survival of the parents' genes more likely. Evolutionary theory allows, however, at least two other exceptions. First, the beneficiaries of helpful behaviour may be close relatives other than offspring. Second, the beneficiaries may be only distantly related, provided that they may be expected to give help in return.

One thing the theory does not allow is selfless conduct which tends to help the species or the group but endangers the individual. In the past, it has often been thought that some animal behaviour exists because it favours survival of the species. Most evolutionists doubt, on theoretical grounds, whether such 'group selection' is possible, except in very unusual circumstances. Few investigations of actual behaviour have suggested anything to the contrary.

Much of the controversy about *The Selfish Gene* and similar writings has, however, been aroused by their supposed implications for humanity. Some writers seem to state that we are forced by our evolution and by our genes to behave in certain ways. But we do not know how apes of a million years ago evolved into human beings; and we certainly do not know what aspects of their social action were crucial for the survival of our ancestors. What we do know is that human beings are often unselfish and have the ability, by taking thought, to change their own behaviour: we can make intelligent and moral choices.

The Origin of Species?

Our account of evolution has to cover not only progressive changes on single lines of descent: it should also deal with the splitting of one line into two. How do two new species arise from a single ancestor? In spite of the title of his most famous work, *On the Origin of Species*, Darwin did not solve this problem.

To begin, we need a generally accepted way of deciding what is a species. People usually identify species by looking at them. Agreement is then often surprisingly easy. Tribal peoples make the same distinctions as do biologists: a remote group in New Guinea has names for 174 species of vertebrates; all but four are identical with those listed by European classifiers. Another method is based on reproduction. Suppose you are wondering whether the horse and the ass, which are very similar, should be classified together. You ask whether they can interbreed and, if so, whether they produce fertile offspring. The mules derived from crossing horse with ass are sterile. So, by this criterion, horse and ass belong to separate species. Similarly, crossing lions with tigers in zoos has occasionally produced tigons or ligers but they too are sterile.

However we define (or fail to define) species, most organisms clearly fall into kinds sharply separate from closely similar types and do not breed with them. How did they originate? Since 1929, we have known one answer. The story concerns two plants of the genus *Primula*, familiar to gardeners: *P. floribunda* from the Himalaya and *P. verticillata* from southern Arabia. They can be crossed. The hybrid is sterile but, unlike a mule, it can be grown from cuttings. A major mutation then sometimes occurs: the number of chromosomes doubles, from eighteen to thirty-six. The resulting plants are fertile and make a new species (named *Primula kewensis* from the famous Kew Gardens in outer London).

Some plant species have been deliberately created and are economically important. Wheat (*Triticum*) has been crossed with rye (*Secale*) to give 'triticale'. This is a sterile hybrid but its chromosome number can be doubled by a chemical treatment. The result (*Triticosecale*) is fertile; it also gives the high yield of wheat, has the rugged growth of rye and is even resistant to a destructive fungus, wheat rust. Perhaps many species have arisen in a similar way in nature. The chromosome numbers of closely related animal species are often different; hence evolutionary change must have involved large alterations in genomes.

Much evolution is nonetheless assumed to be the outcome of changes in individual genes. Some genes act early in development and have large effects on structure. We know this from studying mutants. Flies (Diptera), unlike most insects, have only one pair of wings: instead of a second pair, they have small, club-shaped organs, called halteres, which regulate balance. In Permian rocks, of about 250 million years ago, fossil dipterans have been found with two pairs of wings. The capacity to produce a second pair is still present in modern flies (compare the teeth of birds, page 25). A mutant gene of fruit flies makes wings grow instead of halteres—a return to the ancestral condition. Perhaps a mutation in the opposite direction caused the change to one pair of wings. But such helpful mutations can have turned up only rarely: most mutant genes which cause large changes in development are killers.

Much evolutionary change is therefore believed to be due to accumulation of small changes. If so, how can such a process lead to separate forms which do not interbreed? Probably, geographical separation has often been crucial. Its effects are seen on islands. Hawaii makes an extreme case. Its four large, mountainous islands and many smaller ones are separated from the nearest continent by nearly 5000 kilometres. They are, or were recently, the home of thirty species of brilliantly and diversely coloured honeycreepers and sicklebills (family Drepanididae). All seem to have evolved from a single finchlike ancestral form. A general feature of finches, throughout the world, is a short, strong beak suitable for a tough diet of seeds. Some of the Hawaiian species do eat seeds, but most have quite different diets. Some specialise on fruit, others on insects. One extracts beetles from wood like a conventional woodpecker (family Picidae). An extreme of specialisation is that of honeycreepers whose bills exactly match the tubular flowers of the genus *Hibiscadelphus*. These birds are becoming extinct and the plants they pollinate with them.

Evidently, millions of years ago, members of a single species survived the long journey from the mainland; and its descendants, isolated on the separate islands, evolved special features. No eaters of insects or of nectar were yet present on this remote archipelago. These ecological niches were therefore occupied by the only small land birds present.

The same applies to Darwin's finches. During her famous voyage, *Beagle* visited the Pacific Galápagos Islands, more than 900 kilometres off the coast of Ecuador. Thirteen species of finches (Fringillidae) live there. Although they are quite boring to look at and their songs are unmusical, biologically they are fascinating. They are identified as members of the finch family by details of internal structure, for their beaks and feeding habits vary greatly: one species has the habits of a woodpecker; another eats leaves; a third is a typical insect eater with a long, fine bill; and a fourth resembles a warbler. An oddity of history is that Darwin did not realise their importance until long after his return home.

If many species have evolved gradually by small differences, we should be able to find examples of incomplete separation. And so we can. The herring gull (*Larus argentatus*) and the lesser black-backed gull (*L. ridibundus*) are common in western Europe. We now move west. The herring gulls in North America are slightly different from those in Europe but are still easily identified as herring gulls. Further on, in Siberia, we find more differences: there the gulls begin to look rather like lesser black-backs; and further still, in Russia, the resemblance is still greater. Finally, we arrive back in Europe, which is shared by the two species, now distinct. Hence these gulls form a ring around the northern hemisphere; and a series of transitional forms between the two species remains, spread over this vast area.

Species can, however, probably arise without distant separation and without anything as drastic as a large change in chromosomes. Slag heaps, thrown up as waste from mines, provide an environment which contains unusually large traces of tin, lead and copper and has been investigated by an English botanist, A.D. Bradshaw, and his colleagues. The metals poison most plants but at least one grass, *Agrostis tenuis*, survives. The resistant variety is genetically different from ordinary grass of the same species; it also comes into flower, on average, at a time different from the usual. It is therefore beginning to be reproductively isolated from the typical form. Evidently, this species is in the process of dividing into two. No doubt the botanists concerned are as anxious to preserve the slag heaps as are others to protect forests: they need more time for their researches.

Is Everything for the Best?

Every existing species is presumed to be the outcome of a prolonged process of selection. This (we may think) should, after all this time, have produced the best possible results. Darwin himself wrote of evolutionary change tending toward perfection. If each species is precisely adapted to its environment, departure from the typical form should be disadvantageous. Correspondingly, the effects of mutation are usually unfavourable.

As we know, it is tempting to think of an organism as a mechanism precisely designed for a purpose, like an aircraft or a computer. The facts, however, present a different picture. To give a glimpse of it, here is a series of questions about evolution. The answers reveal organisms as marvellous but imperfect and adapted to their environments often by systems of bizarre complexity.

Are organisms perfectly adapted? Close scrutiny of organisms often reveals features which, in a machine, would represent an engineer's blunders. We can see this in ourselves. Our eyes are impressive organs which enable us to do remarkable things, but they do not transmit light quite accurately and they are slightly (sometimes seriously) misshapen. H.L.F. von Helmholtz (1821–1894), the great German physiologist, said that a human designer of optical equipment would be ashamed of them. Our vertebral column and attached muscles (which are of inconvenient complexity) are evolved from those of quadrupeds whose backbones were supported at both ends. We are upright and many of us have backache. Moreover, our vertebral column includes intervertebral discs which are vestiges of the notochord of our remote chordate predecessors; the pressures to which they are subjected sometimes lead to prolapse of a disc and crippling pain. Worse, our upright posture requires narrowing of the bony pelvis and the difficulties and pains of childbirth. At the chemical level we have a marvellous immune system which protects us against infection; yet this too can go disastrously wrong: the result may be fatal autoimmune disease (such as myasthenia gravis) in which the system treats the body's own proteins as foreign.

If evolutionary change is a result of small alterations in what previously existed, such imperfections should not surprise us. Moreover, evolution is not only in the past: it is still going on. It entails a continuous process of adjustment to circumstances. This can never be complete, if only because the conditions in which organisms live are always changing.

Can we always identify the function of a trait? In biology it is usual and reasonable to ask of any character, what is its function? That is, how does it help survival or reproduction? Sometimes the answer seems obvious. Strange but clearly effective features include an orchid's flower, a chameleon's colour changes and a spider's web.

But it is rash to jump to conclusions. In traditional Darwinian stories, the giraffe's neck is prominent. Its inordinate length enables its possessor to feed on nourishing leaves out of reach of other, less well endowed mammals. Should we then conclude that it is a result of 'selection' to enable it to feed in this way? Female giraffes are shorter than males of their species yet they contrive to feed just as well. Neck length, however, accurately matches leg length, and enables a giraffe to drink without crouching down. So the neck of a giraffe is perhaps a necessary accompaniment of long legs. The legs presumably evolved as a means of getting about quickly on the African veldt. But we are obliged to hesitate even about this: other mammals with hooves manage without them.

The giraffe also presents another problem. In Kipling's *Just So Stories*, its blotchy pattern is said to be camouflage and to confer protection against predators, especially leopards. But its primary function may be as a social signal, for many visible patterns provide information to other members of the owner's species. Especially with animals more manageable than giraffes, experiments can be carried out to give us evidence about the uses of such traits today.

Without such evidence, we can of course make guesses, sometimes plausible, about how they evolved in the past. Such guesses are especially popular with people who wish to explain human action biologically. The trouble is, with ingenuity almost any trait can be accounted for in this way. I have often wondered why I, like many others, develop pains or malaise in certain kinds of weather. I have also thought of a 'Darwinian' explanation. The reader may care to pause and to devise one . . . The discomfort,

I suggest, is a warning of an impending change. Such an ability could have helped our food gathering or farming ancestors.

My suggestion may seem merely frivolous. Others, however, based on a similar logic, have been worked out in great detail. Food aversions and rejections early in pregnancy (including 'morning sickness') are usually regarded as only a tiresome affliction; but perhaps they were beneficial before agriculture was invented. Gatherer hunters eat a great variety of plants, some of which contain possible toxins. Did the changes in attitudes to food during pregnancy protect the developing fetus? Such proposals have been referred to contemptuously as unscientific games. Sometimes, indeed, they are no more. They can, however, encourage further investigation and new understanding of what is going on in the present. They also lead to another question.

Are all features a direct result of natural selection? Earlier paragraphs show that the design of an organism may be imperfect: hence we may have difficulty in deciding why an organ has a particular form. We are then led to a further possibility: that some features are not in themselves advantageous but are only by-products of selection. This notion was first put forward by Darwin himself. Yet prominent evolutionists have, since then, held out natural selection as all the explanation needed for every feature of every organism—including the human species. Such ultradarwinism seems to adopt natural selection as a transcendental principle like that of a divine Creator. It disregards the imperfections we find in the 'design' of organisms.

Land vertebrates, such as ourselves, have obviously odd structures. Some are mentioned above. Another is the arrangement by which air on its way to the lungs, and food on its way to the stomach, both pass through the pharynx. As a result, we and other mammals cannot swallow food or water while breathing. No engineer would tolerate such an absurdity. It can hardly be supposed to be, or to have been, in itself advantageous. It seems to be an outcome of the position, above the mouth, of the olfactory organ of the fish which were our remote ancestors. Such peculiarities evidently result from the gradual character of evolutionary change. Each step builds on an existing arrangement. The resulting adjustments are often extremely complex. The complexity is especially well seen in embryonic development (chapter 6).

In addition, natural selection, we know, produces multiple effects. When one organ or function alters, so do others. The accompanying changes, or spinoffs, may be without survival value at first, but they may later adapt an organism for a new development. An American anthropologist, W.H. Calvin, suggests that human skills in language, music and tool making all stem from selection for the ability to plan ahead; and this in turn, he says, could have arisen from the need to throw missiles accurately. The ability to make a series of movements, planned in advance, depends on a small region of the prefrontal cortex. A normal person, whose arm is under bedclothes, asked to raise an arm, can do so; but, if this part of the brain is damaged, the ability to plan the simple sequence, of first bringing the arm out and then raising it, is lost. Even if Calvin's idea seems fantastic, the principle behind it is valid. Many traits have evidently evolved, in advance of becoming useful, as accessories of other, immediately advantageous features: they are then examples of preadaptation—a notion of likely significance in explaining many features of human action, including our ability to drive cars, to play chess and to argue about evolution.

Summing up: does evolutionary theory predict? In chapter 2 I give evidence for saying that evolution has happened. The present chapter gives still more but is primarily about the distinct subject of the *causes* of evolution. The concept of natural selection is the only credible explanation we have of evolutionary change. What is its scientific status?

The theories of science often tell us what to expect. Prediction may be very accurate: examples are the times of eclipses, the trajectories of objects such as artificial satellites and the amount of electrical power generated by a turbine. In biology, genetics perhaps resembles this aspect of the physical sciences. Evolutionary theory is different. The achievement of Darwin and Wallace was to *uncover* the process which has, apparently, led to the immense diversity of living things. Amazingly, they did this without knowing anything of genetics or of the biochemistry of heredity. Today, biochemistry, genetics and ecology combine with paleontology to give us a moving picture, lasting for more than 4000 million years, of an impersonal, inexorable process of change in populations of living things.

In our descriptions of this process, the conventional metaphors of struggle or competition are highly misleading. Even the word selection is a metaphor. Natural selection is a consequence of the way in which organisms vary and reproduce themselves. It is the outcome of the remarkable properties of the nucleic acids, DNA and RNA: they promote the synthesis of very complex organic molecules, the proteins, from simpler substances taken from the environment. Although DNA is as a rule accurately copied when a nucleus divides, the copying can fail: mutation is a constant source of variety. Sexual reproduction enhances variation; and diversity is further increased by the formation of many-celled structures: this requires an elaborate development from egg to adult and has resulted in the evolution of large and complex organisms.

The statements in the preceding paragraph are descriptions. Although they summarise fundamental discoveries concerning the nature of living things, they do not enable us to foretell the future. If modern knowledge of evolution had been available a few decades ago, it would still not have made possible confident prediction of the replacement of speckled moths by black ones: the moths might have evolved false warning colours; or they might have been wiped out. And that is a simple case.

We can predict that changes will happen. The transformations revealed by modern biology are, however, not directed by any overriding agency. Biologists can no longer present organisms as the product of an intelligent designer. Only today are we ourselves transcending time and chance and making plans for the future. Hence the survival of many species, including our own, during the next few centuries, will depend on decisions made by ourselves and by our immediate descendants.

Which of these two individuals should we study if we wish to understand humanity?

The Human Species

Our own species, *Homo sapiens*, is a product of evolution. A description is gradually emerging of how we arose from early apes. But to make sense of humanity, even imperfectly, we must realise that, biologically, the human species does not make sense. If we wish to understand ourselves, we need not only human biology but also kinds of knowledge which fall outside the natural sciences.

Chapter 11

Human Nature

> The scope of human instinctive knowledge is limited, and we feel the lack of it. Free will is a hard burden to bear.
>
> J.B.S. Haldane

ZOOLOGISTS FROM A DISTANT planet would, like us, presumably classify the human species within a small group of large animals, the Hominoidea. Our close evolutionary relationship with the apes has been emphasised by modern biochemistry. Nearly 99 per cent of the DNA in the nuclei of human cells is identical with that of chimpanzees. Some writers, who hold out the apes as models of humanity or as sources of knowledge of human society, present this chemical likeness as telling us something very significant about ourselves. They could take this reductionist argument a step further. Dehydrated and reduced to chemical elements, a human body consists of carbon, hydrogen, oxygen, nitrogen, iron, chlorine, phosphorus and others, in precise proportions. So does the body of a chimpanzee, in much the same proportions. It does not, however, follow that, faced with the problems of human society, we should journey to central African forests and acquire wisdom by observing *Pan troglodytes* or *P. paniscus*.

The living hominoidea

Lesser apes (Family: Hylobatidae)
Gibbons (*Hylobates*); several species. South East Asia.
Siamang (*Symphalangus syndactylus*). Malaysia, Sumatra.

Great apes (Family: Pongidae)
Orang utan (*Pongo pygmaeus*). Borneo and Sumatra.
Chimpanzee and bonobo (*Pan*); two species. Tropical Africa.
Gorilla (*Gorilla gorilla*). Tropical Africa.

The human species (Family: Hominidae)
Human beings (*Homo sapiens*). Everywhere.

The human species is sometimes put in the same family as the apes; and the siamang may be treated as a species of Hylobates.

THE HUMAN ANIMAL

Instead we now inspect the human species itself, seen through the eyes of intelligent, extraterrestrial visitors. Many of the visitors' observations are on people before agriculture was invented, that is, on gatherer hunters. They notice at once that human beings move readily on the ground on two legs only. The hind feet are not used for grasping objects but have an arch which adapts them for running. The forelimbs are short. The shoulder joint is fabulously mobile and makes possible many movements such as those of throwing and bowling as well as climbing and swimming the crawl. The forefeet have become hands and the opposable thumb provides a strong grip. Objects, such as stones, sticks and bones, are with great dexterity fashioned into tools. All groups of this species depend on manufactured equipment for their survival.

The human head is large but the skull bones are light. The canine teeth do not form conspicuous fangs. The teeth are smaller than those of apes but can masticate tough plant food such as seeds. *Homo sapiens* is a versatile feeder and can also digest meat. Correspondingly, human beings have no single ecological niche and have spread into an astonishing range of habitats.

Most of the body, except the pubic region, has only fine hairs. The head, however, has much hair, sometimes long, sometimes short and curly. Adult males usually have prominent hair on the face and some have it on the chest and belly as well.

The skin is often dark, even black, owing to the presence of a pigment, melanin; but, especially in cold regions, there may be little skin pigment. Often, the skin is kept clean by washing. Pheromones secreted by the skin are then largely removed; sometimes they are also disguised by odorous substances. Much of the body surface may be covered with clothes which are sometimes brightly coloured. Metal and other embellishments may be dangled from the neck or attached to other parts of the body. Great variation exists in all these customs.

Muscles allow many changes of facial expression. The eyebrows can be moved into a variety of positions which reflect different attitudes and emotions. The eyes have whites (the sclerae) which make it easy to see what a person is looking at. Lips with pink rims make the movements of the mouth easily visible. All these features assist communication.

Information, however, is transmitted mainly by sounds which are far more elaborate than those of other species. More than 30 individual sounds, mainly grunts, cries and clicks, are produced at rates of up to 25 a second and are strung together to make words and sentences. Parts of the brain especially related to movements of the larynx, that is, to speech, are very large; the human larynx has a structure different from that of apes.

The brain is enormous and out of proportion to the rest of the body. At 1300 to 1400 cubic centimetres, it has about three times the volume of the brain of an ape of similar size. Some regions, concerned with understanding as well as producing speech, hardly exist in other species.

Human reproduction too has special features. The penis is unusually large but the testes are rather small: even if, exceptionally, a male achieves several ejaculations in a day, only one is likely to contain many sperms. In contrast, a chimpanzee can copulate once an hour and can produce an ejaculate with a high sperm count four times in a day. Male Primates usually copulate from behind the female. The bonobo, however, often copulates face to face. The large human penis allows coitus in several positions; face-to-face attitudes are perhaps the most usual.

From the age of about thirteen to around fifty, unless pregnant or lactating, a female ovulates every 28 days or so. Many other Primates have a coloured sexual skin which surrounds the genitalia and becomes conspicuous when the female is receptive. Human beings have none; and, although changes of mood often go with the menstrual cycle, coitus may take place at any time. This condition of 'permanent estrus' is puzzling: coitus must often occur when pregnancy cannot follow, which seems wasteful. Perhaps, however, it promotes survival by keeping mated pairs together.

The mammary glands of other species begin to enlarge shortly before the birth of the young, when milk will be needed. Human breasts enlarge at puberty and remain prominent. The enlargement is due to fatty tissue: size is not related to milk production. Adult females usually accumulate adipose tissue also on the buttocks and hips. The resulting bodily contours have been thought to attract males and to confer a biological advantage on their possessors. But males of different groups (and even within groups) differ in their preferences: some seem attracted by an almost fat free (boyish?) figure. The prominent facial hair of the male has similarly been held to attract females but the widespread habit of shaving does not seem to reduce fertility. In all this, as in care of the body surface, local fashion (the cultural influence) is evidently important.

Each species of ape has its own distinctive mating pattern. Gibbons pair off (monogamy) and live in family groups with several young. Orang males have a territory which includes the ranges of several females (polygyny) but they are said not to copulate very often. Male gorillas too are polygynous and infrequent copulators. Chimpanzee males are promiscuous (polybrachygamy) and copulate frequently; and, although consort relationships are common, the females accept many males (polyandry). In contrast, the single species, *Homo sapiens*, includes men with several wives, women with several husbands, monogamous pairs, promiscuity and even group marriage—another example of human diversity and adaptability. Yet marriage, in the sense of prolonged association of a female with a male, is widespread.

The young are born in a helpless state and depend on parental care for many years. The methods of rearing young vary greatly. A unique feature is that older individuals (not only parents) systematically teach the young a variety of skills.

FRAGMENTARY FOSSILS

How did this strange species evolve? Our visitors from space do not know. Nor do I. The information available consists largely of fossil bones, some of them only lower jaws, with or without teeth. A single jaw, even a tooth, may be given a long name and have an elaborate story built around it. Occasionally a fossil is given a simple name as well, such as Lucy.

The enthusiasm of paleontologists has led, during the past century and a half, to much speculation. The most famous example is an early one. In 1856, in Germany, an unusually complete fossil human skeleton was discovered and called Neanderthal Man; later, similar fossils were found in many other places. The first descriptions were of an incompletely upright being, bowed rather in the manner of an ape walking. Here, it was thought, was a link between the apes and ourselves. So artists made imaginative reconstructions and gave them brutal faces. But, in 1957, two English workers showed (what had already been suspected) that the person concerned had severe vertebral osteoarthritis: he probably went about, as some of us do today, rather bent, holding his back and perhaps even chewing an analgesic plant recommended by a witchdoctor. Despite his age and his back, he may have been quite good-looking.

Enough of fantasy. For more than twenty million years, a variety of apes lived mainly in forests. Most of them probably weighed less than 30 kilograms (the weight of a European child of about ten years) but one line led to a creature larger than the largest gorillas. Its name, *Gigantopithecus*, was first applied to teeth discovered in Chinese drugstores: they were called dragon's teeth and sold for medical (or magical) use. Later, lower jaws were dug up in China and India. *Gigantopithecus* is thought to have lived mainly on the ground, in open country, and to have eaten a variety of plant foods. It is, I fear, unlikely that a few remain in the Himalaya as the source of stories about the yeti, or abominable snowman.

Now imagine a paleontologist faced with the fossilised bones, without the skull, of a newly discovered type. The bones are dated three or four million years ago. They seem to belong to a rather small, rather odd human being: evidently, the creature walked erect; the feet are arched and the hands have grasping thumbs. But later a skull becomes available too, found in conditions which make it certain that it belonged to the same individual. And the skull is *not* like ours. Although the brain is larger than that of the apes just described, its cranial capacity, of about 600 cubic centimetres, is still less than half that of a typical human being. The jaws and teeth are rather human, but they are more massive and protrude like those of an ape (prognathism). The cranium is heavy and has large brow ridges above the eyes, again like those of an ape.

Many such fossils have now been discovered. They may be put in several species of a single genus, *Australopithecus*. The earliest known, *A. ramidus*, (also called *Ardipethecus ramidus*), from Ethiopian rocks about 4.5 million years old, was first described in 1994. It lived in woodlands and probably walked upright. The males, at about 1.7 metres, were the height of a very short man; the females, at one metre, were much shorter. A later and better known form, *A. afarensis* (which includes Lucy), of about three million years ago, was certainly upright; we even have footprints of three of them, preserved in hardened volcanic ash.

So the erect posture arose before the typical human skull and brain. We do not, however, know how the various australopithecines lived. Gorillas and the two extant species of chimpanzee have quite different social lives. Each extinct species, too, presumably had its own distinct way of living.

Nor do we know what impelled the striking changes from ape to humanity. A recent study, aided by computerised models, shows how difficult it is to do more than guess about evolutionary changes. During exposure to the tropical sun, an upright posture has been found to reduce the need for water. (Recall the importance of sweating.) Perhaps selection for bipedality has this prosaic, physiological explanation.

Another obstacle to understanding is the imagination of some writers. One widely propagated story presents these precursors of humanity as bloodthirsty big-game hunters violent also among themselves. In fact, however, serious study of the remains shows typical australopithecines to have been not predators but prey: they were evidently hunted by large carnivores.

The australopithecines make a series of 'missing links', no longer missing, between extinct apes and the earliest beings we call *Homo*. The most important species of our genus (apart from ourselves) was *H. erectus* (formerly *Pithecanthropus*). It arose about two million years ago and survived for more than a million years—longer than we have, so far. It was larger than most of its predecessors; the skull bones were still thick but the face and teeth were smaller. The brain was nearly as large as ours. The stone tools were beautifully formed. And *H. erectus* had fire; hence food could be cooked and the dietary range was widened. Correspondingly, the shape of the thorax indicates an intestinal tract smaller than that of apes and similar to ours.

The increase in brain size, after the arrival of *Australopithecus*, was rapid. It went with a progressive lightening of the bones, especially of the skull, and of the teeth. The swollen brain is often assumed to be related to two novelties: making and using increasingly complex tools and making and using fire. Both require planning ahead and both entail transmission by one generation to the next. But, as usual, we cannot be sure.

Homo erectus evidently originated in Africa; but fossils of this type—the famous Java Man and Pekin Man—were first discovered in Eastern Asia. They were the outcome of an extraordinary dispersion, from Africa over most of Asia and Europe: a long walk.

Since the time of *H. erectus*, the Neanderthals have been prominent. They too occupied great areas of Africa, Europe and Asia. Their brains were at least as large as ours. They were heavily built and seem to have ranged widely in the search for food. Most had a distinctive equipment of tools of the Mousterian culture. There is, however, no exact parallel of types of stone industry with the different kinds of fossil human being: some Neanderthals used tools of kinds usually associated with *H. sapiens*. Perhaps this was a result of copying: we do not know how members of the two types interacted. Although Neanderthals are thought to have buried their dead, complete with tools, as if preparing for an after life, the remains regarded as graves may be only the result of falls of rock. It is also uncertain whether Neanderthals had speech.

Two English anthropologists, William Straus and A.J.E. Cave, have nonetheless suggested that, shaved, washed and suitably dressed, a Neanderthal would attract little attention in a city crowd. They may be right; but Neanderthals differed from us in their more massive skulls and teeth and in the absence of a chin; in addition, the back of the skull (the occipital region) was expanded into a large bulge. Even in a dense crowd, they would have looked rather odd. And a very recent triumph of DNA analysis, applied to a Neanderthal bone, seems to have ruled out Neanderthals as a variety of early *sapiens*: they evidently qualify as a separate species of *Homo*, now (regrettably?) extinct.

For tens of thousands of years, Neanderthals and *Homo sapiens* shared the great land masses of Africa, Europe and Asia. Then, a little more than 30 000 years ago, Neanderthals disappear from the fossil record and are replaced, in Europe, by 'Cro-Magnon man'. These newcomers were beings like ourselves, but were on average taller than modern *H. sapiens*. They were also innovative tool makers: their equipment included skilfully made spears of antler or bone. The extinction of the Neanderthals has not been convincingly explained. We have no evidence of violent conflict.

The names given to fossil types, especially in a short account, can easily give a wrong impression: we may seem to know only a number of sharply distinct kinds or species of almost human beings. But we have no good reason to suppose that humanity evolved by large, abrupt changes. As new fossils are accumulated, the separation of the various named types becomes increasingly blurred. The gradual evolution of the human species seems to have happened, like that of earlier forms, in Africa. *Homo sapiens*, like their probable ancestors, *H. erectus*, then evidently made another long march over much of the rest of the world.

All members of the genus *Homo* made stone tools. We have a continuous record of such artefacts from about 2.4 million years ago. The earliest were pebble tools (Oldowan), some of which are associated with skeletons variously assigned to *Australopithecus* or *Homo*. They are chunks of rock chipped at one side to make a sharp edge. Though crude, they are products of skill: making them must have required much practise. Pebble tools were later chipped on both sides and sometimes all round. Hence the beings responsible had already gone well beyond anything achieved by chimpanzees. They sometimes carried chunks of rock more than ten kilometres: great concentrations of tools and fragments have been found at their centres of activity—precursors of our manufacturing industries.

The next phase began gradually, about 1.5 million years ago, and was that of the hand ax, made by more elaborate chipping of larger chunks of rock. A later kind was minutely chipped over the whole of its surface and formed into the distinctive, standard shape called Acheulian. *Homo erectus* (or close relatives) turned out these objects for more than a million years.

But were they axes? Again a traditional story has been doubted. Hand axes are usually described as core tools; but the flakes may have been more important than the cores and were possibly used for scraping or cutting. Or perhaps the cores were used as missiles: these 'axes' are flat and can be thrown with a low trajectory, like a frisbee.

With the arrival of *Homo sapiens*, the small flakes struck from the cores were modified by grinding as well as chipping; they were shaped with great skill and some were used as knives or points. These people also made equipment, already mentioned, from bone, ivory and antlers. They used heat to alter the properties of materials and invented spear throwers and bows and arrows. Stone Age people have, for at least 40 000 years, also been artists: they have painted pictures, some brilliant and dramatic, on rock surfaces (famously in caves) and on bark; they have carved designs on bone and some of their stone tools seem to have been works of art too. For the same period they have used beads and pendants as ornaments. They were people with whom we can easily imagine having a friendly understanding.

Impostors: 'Cave Men' and their Successors

We do not know how all these beings, from *Australopithecus* on, acquired their skills. Even if the first manufactured tools were inventions, the methods of making them were presumably passed on as traditions. Today our ignorance of Stone Age society is acknowledged. Until recently, however, 'primitive man' was seen as a hunter, probably persistently violent both to big game and to his fellows, and perhaps also to his women. Men were supposed to provide food and to protect their families from danger. Women and children stayed at home (believed to be a cave). Stone tools were thought of primarily as weapons. The authors of such stories were all men and European. At the time, their countrymen were invading and taking the land, the women and the lives of less well equipped people. They were also using their guns to slaughter big game. The pioneers often resented the sometimes hostile response of the 'natives' and attributed it to primitive depravity.

Today, anthropologists are far more restrained. They have much more information about early human beings. Many more fossils have been found. To identify the foods they once chewed, fossil teeth have been minutely examined under the microscope. Stone tools, similarly inspected, have revealed the uses to which they were put. Accumulations of bones, closely studied, have provided evidence on whether they were the result of scavenging or of hunting. Anthropologists have lived with some of the remaining gatherer hunters, especially in Australia and the African Kalahari; and, with proper humility, they have learned some of their impressive skills (such as the construction and use of stone tools) and elaborate languages. Yet others, in Africa, have put themselves in the position of gatherer hunters who wanted meat and have found the best source to be dead or dying animals.

From all this information, we find small, mainly peaceful groups, living in the open, not caves, and gathering much plant food. The animals caught (often by women and children) are usually small—for instance, lizards, frogs and insects. Meat may be acquired from large animals by scavenging. Big game hunting is unusual. Tools are used for cutting plant tissues, for whittling wood and for chopping up animals: they are prototypes of kitchen and garden equipment. Fire is mainly for cooking.

Such people are without personal property as we know it. Cooperation among members of a group, both in getting food and in caring for the young, is crucial for survival. Surviving gatherer hunters also have very complex languages, used not only for teaching the young skills and proper conduct but also for telling elaborate stories about natural phenomena, such as thunder, rain, wind and floods, and animals. As well, gatherer hunters have time for painting, sculpture, ornament and music.

In the middle of the twentieth century, widely read popularisers have nonetheless revived images rather similar to that of the cave man. They are based on the presumption that studying the social lives of other species can tell us what to expect of humanity. This is the fallacy of biological naturalism, already encountered in chapters 7 and 9. It is a fallacy if only because the behaviour of one animal species does not reliably show even what we can expect of other animals. Ideas about human beings derived from the animal kingdom, to be of use, must be tested by observation of ourselves (compare pages 185–6, on the effects of crowding). On our intellect and our morals, animals tell us little or nothing.

One modern successor to the 'cave man' presents us as *Homo pugnax*, a species whose members are implacably hostile to their fellows. According to surveys in eighteen countries, around fifty per cent of people believe that 'the roots of war are in man's basic nature'. This belief is propped up by errors in apparently authoritative writings. 'Man' has been described as a hunter and therefore as a fighter. Many animals, such as tigers, wrens and spiders, do kill and eat animals *of other species*. These predators, however, do not systematically hunt down their own kind. And plant eaters are neither more nor less violent among themselves than are the meat eaters. Much intolerant animal behaviour is harmless, but some animals do attack and injure members of their own species. What makes them do so has to be discovered by observing each species separately.

A persistent notion is of a biologically fixed drive or instinct for aggression which forces us all to be violent and even to make war. Yet most people are peaceable nearly all the time. The decision to go to war is not forced on us by our biology: it is taken by elderly persons after calculation. War fever among the populace is worked up later. Soldiers have to be trained to fight: impulses of friendship, tolerance or helpfulness have to be suppressed and hatred encouraged. These efforts are not always successful.

Some kinds of individual violence ('aggression')

Assassination for pay
Murder of a public figure by an insane person
Hanging a hated person during a race riot
Rape
Wife beating
Clitoridectomy
Thrashing a child for wrongdoing
Hitting a parent during a temper tantrum
Injuring an opponent during boxing or football

Above are a few of the items sometimes put under the heading of human aggression. To understand them, they must be separately studied by the methods of social science, psychology, education and medicine.

Homo pugnax is one of the portrayals of humanity by which, throughout history, people have expressed gloom about themselves or contempt for others. The latest of such images is owed to human sociobiology. Here is a list of unwelcome human attitudes with their opposites.

	Contrasted attitudes
Violence, lust	Restraint, self control
Hostility	Friendliness, love
Servility	Independence
Greed	Generosity, obligation
Selfishness	Altruism, disinterestedness
Nepotism	Impartiality
Deceit	Truthfulness
Illogicality	Rationality, objectivity

Sociobiologists attribute those on the left, especially selfishness, lust, deceit and nepotism, to genes and natural selection. They are therefore said to be part of our biological nature. We are also represented as having a relentless drive to breed. Hence we (and the authors of these stories) become *Homo egoisticus*.

The writers who emphasise the wickedness of humanity are themselves human. To repeat what I have written elsewhere, none has (I believe) said anything like this:

> I am aggressive, competitive, selfish, spiteful, deceitful and ruthlessly bent on contributing largely to the next generation: I therefore maintain a harem.

They do, however, state that sociobiology can show how human societies adapt themselves to particular environments and so increase the 'genetic fitness' of the group. But they do not publish measures of fitness. Instead, their revelations about their own species include the following: that people often defend a region from other people; parents look after their own children in preference to those of others; relatives help each other; men are attracted to the female form and women to the male; human beings, though highly social and sociable, are sometimes violent, false, selfish and greedy.

Hence they make lists of things that we know we do, often

with special emphasis on ill doing. They then say that they have explained them. They also contradict themselves. E.O. Wilson, the leader in this field, has stated, in total opposition to his most prominent message, that almost all differences among societies ('cultures') are based on learning and socialisation rather than on genes. Such incoherence runs through all his writings and those of his followers.

It has therefore often been asked: what new, important discovery, about human social life, has come from sociobiology? The rest was silence.

The Talkative, Calculating, Singing, Teaching Species

A philosopher, Ludwig Wittgenstein (1889–1951), once commented on the invisibility of important things: we cannot notice them, he says, because they are always in front of our eyes. So the next paragraphs describe further, in his phrase, some of the 'simple and familiar' features of our species—features which are often concealed by the impostors.

Our animal ancestors must have exchanged news by grunts, cries, gestures, odours and embraces, such as those of animals described in chapter 9. At a time between the coming of *Australopithecus* and the rise of *Homo sapiens*, language evolved. We do not know how. Our languages today are fundamentally different from the signals of animals. Chimpanzees and other Primates make more than thirty distinct sounds, a number similar to ours. Each of these signals refers to the total situation at the time the signal is made. The East African vervet monkey (*Cercopithecus aethiops*) makes a distinctive call on the appearance of a python; another call indicates a leopard. Both animals are dangerous predators. Other members of the troop take appropriate cover. In contrast, we put our separate sounds together to make words. Our word, python, can be used in a great variety of sentences and situations: for instance, 'what is a python?'; 'the python is going away'; 'there is no such thing as a python!'; the 'py' in 'python' does not mean 'pie'. It is easy to think of many others; some could refer to occurrences at a different time or place or to imagined events.

Attempts have nonetheless been made to teach apes to

communicate as we do. Because the structure of their larynx prevents them from making human sounds, signs designed for deaf people have been used. This brilliant suggestion led to many years of dedicated study. As a result, young chimpanzees have been taught to signal sentences such as 'please give-me orange'. Some chimpanzees, taught to sign by experimenters, have taken to making similar signs among themselves. Bonobos have achieved more. A male, Kanzi, picked up sign language by observing attempts to teach signs to another ape. Later, two other bonobos and a chimpanzee did the same. Kanzi, by the age of eight years, also learned to respond appropriately to quite complex commands, such as 'Pour the coke into the lemonade'.

These researches have, however, also displayed the gulf between apes and ourselves. A child begins in infancy with babbling and later may invent sounds which resemble words. Soon, real words are uttered. During the second year and later, a child's vocabulary enlarges at an amazing rate: by four years, 10 000 words may be understood and used. (An adult typically has 30–40 000.)

Words are assembled in increasingly elaborate sentences. Quite early sentences contain two distinct clauses: 'I know/she's here'; 'tell me/what to do'. The achievements of children now far surpass those of apes. An early development is the ability to understand and then to use grammatical items, such as 'to' and 'if', which are not represented in the signals of animals. (About half our words are of this kind.) W.H. Calvin gives the example of the nursery rhyme:

> This is the farmer sowing the corn,
> That kept the cock that crowed in the morn,
> . . . That lay in the house that Jack built.

Quite young children understand what is implied by 'that' in the rhyme.

A child acquires the complexities of language by hearing people talking and by taking part in conversation. Any normal human being can, at least when young, learn any language. Learning new words can continue throughout life. How this happens is still obscure: it is not mere parroting. Even a young child can form words into sentences which neither the child nor the surrounding adults have used before. This is the productive quality of language. Often, a young child's sentences are grammatically slightly wrong: for instance, 'I throwed it'. Such errors are gradually corrected, even if nobody points them out. Moreover, a child soon becomes able to describe something that happened, or might happen, at a different time or in a different place. This feature of language is called displacement.

Once learned, language has a number of uses. Consider the possible replies if the reader were asked to comment on this book. The reply could include, first, *indications of feeling*, such as 'What rubbish!'.

A second possibility would be to *describe* the book: its contents, style, illustrations and so on. Unlike animals, people describe things from an early age. Many descriptions (such as those by some book reviewers) are invented. In an additional complication, the ability to describe complex phenomena, like other skills, varies greatly. To give an effective description of a person, a crowd, a flower, a landscape or a book is an art which has to be learned. (The same applies to drawing such objects—another form of description.)

A further comment would be to ask *questions*. For instance, if evolution has been going on all this time, what changes are going to happen next? (I don't know.) Asking questions can lead to disagreement and to the *argumentative* use of language. Perhaps commenting on the book will lead to discussion and debate.

One important kind of description depends on numbers. Although chimpanzees are said to be able to learn to count up to ten, mathematics is confined to the human species. We are the only species able to make a census of a large population of ourselves. Numbering, like speech, though universal takes many forms. Today we are accustomed to standards of measurement accepted throughout the world, such as the metre and the gram. This is a modern development. The English acre was originally the area which could be plowed in one day; it therefore varied from place to place. It is now defined as exactly 0.40468 hectares.

Just as language has developed increasing complexity in the hands of philosophers, poets and the rest of us, so mathematics has become, over thousands of years, in some of its forms more elaborate and more abstract. It is now a massive discipline and is still growing.

An account of language, however brief, ought to include

definition. A reader who asks for a word to be defined probably expects an answer from a dictionary. This is a lexical definition and is a description: it states how a word has been used (and, as a rule, still is used). It can therefore be wrong. (A lady once asked Samuel Johnson why, in his dictionary, he had wrongly defined 'pastern'; he replied, in his usual uncompromising way, 'Ignorance, madam, sheer ignorance'.)

Another use for language, crucial in science, is making stipulative definitions. These are promises or statements of intention. When somebody introduces a new word, such as 'homeostasis' for maintaining a steady internal bodily state, the inventor undertakes to use the word in just that way and no other. Unlike a lexical definition, a stipulative definition can never properly be said to be incorrect.

Most of the statements in the preceding paragraphs apply to all languages. Human speech, however, is not one tongue but many. Each language can change in a few years and always (it seems) does change in a few centuries. Languages are very adaptable. At any time, new words can, like homeostasis, be invented to match new phenomena or ideas: other recent examples are 'gas', 'electricity' and 'genetics'. The scope for novelty in language is infinite.

So language is universal; but, simple or complex, it takes many forms, it has many uses, and it changes, sometimes rapidly, with time. The same applies to music. Presumably music began as songs sung by individuals. It would be interesting to know when duets and choruses arose and why. Later, evidently, people joined in with instruments. Perhaps, when some mesolithic Monteverdi improvised tunes for the first time on a wind instrument, the novel noises were condemned by tribal elders as a newfangled absurdity.

Today, as for thousands of years in the past, music is heard on formal occasions, such as coronations, military parades and religious ceremonies; it accompanies dancing and stage shows and may, as in opera, be an essential part of drama; for at least 2500 years it has been used to treat illness; it is sometimes part of courtship; it sends babies to sleep; above all, it is composed, heard and performed for its own sake. All this is a far cry from signals such as bird song or other animal sounds.

Speech, mathematics and music all depend on transmission from one generation to the next. The transmission is by example and by teaching. In chapter 9, I define teaching as the process by which one individual shows another (the pupil) how something can or should be done and *tries to persist until the pupil achieves a certain standard*. (This is a stipulative definition.) As we know, not even chimpanzees systematically teach skills as we do. We are all teachers from an early age (some more than others). Teaching maintains traditions but also allows rapid changes in knowledge and custom. The changes may occur among people unrelated to the teacher. Such social evolution takes place without any help from natural selection or altered genes. It reminds us of two central components of human life: our need to take decisions on what we ought to do; and our ability, however imperfect, to influence what is going on around us.

Teaching as a feature typical of our species has been neglected by social scientists. Perhaps this is because schools have commonly been places of punishment. This is now changing. Teaching is increasingly recognised as an activity aided by the capacity to imagine the difficulties of others and how others are feeling: that is, by empathy.

A teacher's exertions are also effective only because others, especially the young, have the capacity to profit from them. This ability requires not only intelligence in its everyday meanings: it demands empathy on the part of the pupil, an important phenomenon which psychologists are only now beginning to study. Empathy in turn depends on awareness of one's own feelings. Human beings know not only a great range of phenomena in the outside world: they are aware also of themselves both as individuals and as part of humanity.

The preceding passages illustrate human self consciousness. They may be thought of as summing up human nature. In daily conversation, however, 'human nature' commonly means what the speaker considers inevitable because it is customary in a familiar community. In the past, even highly educated and thoughtful people have considered it natural that some people should be slaves and that women should be totally subordinate to men. Similarly, the impostors described in this chapter imply that our nature makes us incorrigibly violent and motivated by greed or by the impulse to have many children.

Above I describe features, truly common to all *Homo sapiens*, which are not shared with other species. None is a fixed trait in the same sense as is walking upright. Yet they have been called instincts. Today, they are sometimes said to be 'biological'. Such expressions imply that they are somehow fixed (by the genes?). If so, should we add washing, wearing clothes, not wearing clothes, making music or keeping pets? Although these are or, for long periods, have been general features of the daily lives of human beings, perhaps few people would now accept any of them as instincts: for they are obviously not fixed but are greatly influenced by local custom (or culture).

The limitations of talk about instinct appear in our capacity for speech. This universal human trait no doubt reflects a genetical difference of our species from apes. Its important feature is, however, not the fixity implied by saying that it is an instinct or genetically determined: it is the complexity of its development in each individual. The growth of speech and understanding in childhood is marked by immense variation. Imitation plays a part; so do example and encouragement from parents and others. The forms of these social interactions depend on local convention. Even describing, let alone explaining them, is difficult. To say that we have an instinct for language is therefore only a way of saying that linguistic abilities are typical of our species and develop reliably in most environments. The interesting questions concern how the development takes place and how it can be enhanced.

Correspondingly, the human nature presented in this book shows human conduct as much less easy to pin down than is the behaviour of animals or of imaginary cave dwellers and their successors. Other species have fixed social interactions and are usually restricted to narrowly defined conditions, whereas our ways of rearing children and managing our diet and other vital activities are, like our languages, far from standardised. Our social lives and our survival depend on learning skills and on learning and modifying traditions. Today we have to choose what to teach our children about an increasingly complicated and enigmatic future. Among our most crucial decisions are those on our relationships with the living world of which we are a part.

Chapter 12

Humanity in Nature

> Don't panic.
> Douglas Adams

In 1890, a white American quoted the response of a native American to European farming. 'You ask me to plow the ground. Shall I take a knife and tear my mother's bosom?' An Englishman, at about the same date, wrote of the Indian farmer:

> The monkey, the nilghai, the black buck, the wild pig, and the parrakeet fatten at his expense . . . The principle of abstaining from slaughter is pushed to an almost suicidal point in purely Hindu regions.

To Europeans, the attitudes of the native Americans and of the Hindu cultivators were equally strange. The teachings of the Iron Age pastoralists who gave us *Genesis* were more familiar. 'Thou shalt eat the herb of the field; in the sweat of thy face shalt thou eat bread.' The same people were confident of 'dominion over the fish of the sea, and over the fowl of the air, and over the cattle . . .'. Today, however, many in the technically advanced world are coming to treat with greater respect the outlook of people in simple societies. The ethnobotanists, who search the world for valuable species, have met, in almost every indigenous group, the belief that the entire planet is sacred.

The existence of such diverse outlooks reminds us how various are human presumptions about nature, as about everything else. The practices some take for granted today are not fixed. This is fortunate, for in much of the world drastic change is urgently needed.

THE BIOSPHERE ERODED

Now that the human population has reached the stage of doubling in 35 years, the biosphere is changing at an unprecedented rate. Previous chapters describe some of the effects on fisheries, forests and grasslands. Here are some additional facts. When the Portuguese first arrived in Brazil, they found enormous areas of primeval rainforest. Today, only three per cent of that forest remains. Throughout the world, in the 1980s, despite much new sowing, many more trees were cut down than were planted. A cautious estimate was of a loss, every year, of eleven million hectares of forest. Forests are vast consumers of carbon dioxide, which they convert into their tissues; they provide us with a great variety of valuable products in addition to wood; they house a large proportion of the world's animal and plant species; and they are very beautiful. Sometimes, loggers seem to be like members of a family whose only shelter is a handsome timber house and who gradually break it up, room by room, to keep a fire going.

Grasslands or drylands are similarly disappearing. One measure is loss of soil. Soil formation depends on weathering of rocks and on the growth of a succession of plants: barren rock becomes well covered with soil only after decades or centuries. In the 1980s, every year, 25 million tonnes of soil, lost as a result of erosion, was not replaced. (Some estimates are much higher.) In addition, despite irrigation, deserts are steadily encroaching on cultivated land. Humanity is therefore heading for widespread famine of a kind which, as this chapter is written, is confined to parts of Africa.

At the same time, the climate is changing. Much of the visible radiation from the sun is reflected back from the earth in the form of heat (infrared radiation). Some, however, is always retained, as in a greenhouse. An increasing amount of heat now fails to return to space: extra carbon dioxide and other gases, much of them produced by human action, hold it in a trap.

This, usually called the greenhouse effect, is of course an *enhanced* effect. Evidence of its existence has come from analysis of ancient ice in Antarctica, and of the remains of small organisms in sediments at the bottom of seas and lakes. These provide a record covering 160 000 years. During 'ice ages' the concentration of carbon dioxide was about 210 parts per million (ppm); in warmer times (interglacial periods) it was about 280 ppm. Today, the fires of the expanding human population, the chimneys of modern industry and the combustion of petroleum are steadily increasing the amount of carbon dioxide in the air. The figure in 1990 was over 350 p.p.m. Further evidence comes from annual growth rings in the trunks of the long-lived Huon pines (*Dacrydium franklinii*) of Tasmania. These indicate the temperatures in the years in which the rings were formed. Cores can be taken without harming the trees. Rings formed during 1965–1988 are the largest for over a thousand years.

If we turn to the far north, we are reminded that in ecology nothing is simple. Alaska, during the 1980s, had exceptionally high average temperatures and the growth of trees accelerated. But, in the 1990s, growth is slowing. An adverse effect of warming is suspected. Moreover, populations of insects that attack trees have gone up. By 1995, therefore, Alaskan forests seemed to be threatened both by warming and by insects.

As a result of warming, gigantic masses of ice, in the Antarctic, in Greenland and in mountain glaciers, are melting at a rate not equalled for tens of thousands of years. The largest, that of Antarctica, is shrinking rapidly. Water is pouring into the oceans and raising their level. Heavily populated, low-lying regions, especially the Netherlands, Bangladesh, Egypt and oceanic islands, face disaster.

The complexities of the world's water system are only now being analysed. Currents of warm water prevent the poles from being colder than they already are; correspondingly, cold water flows back from them, over the sea bottom, and cools the tropics. (As a result, industrial wastes have been recovered from sediment under the Arctic ice sheet.) If this cooling diminishes, the tropics will become still hotter and crop yields there will decline.

An increase in carbon dioxide concentrations could, at least temporarily, accelerate plant growth. Some of the findings on tree growth, mentioned above, fit this idea. Recent research partly confirms it. The research was provoked by a discrepancy. Laborious calculations had suggested that the rise in atmospheric carbon dioxide was less than expected, by about 2000 million tonnes. Where had it gone? In 1995, an international group used modern devices and computers and spent a year sampling the air above a region of Brazilian rainforest. In the area studied, more carbon dioxide was going into the forest than was coming out. Evidently, one effect of the extra carbon dioxide is to promote the growth of trees in the tropics. The extent to which this compensates for the effects of ozone depletion has still to be discovered. The finding does, however, further emphasise the importance of planting more trees.

With or without extra carbon dioxide, plants can grow well on land only where there is plenty of unpolluted water. Today, rivers and lakes are being filled with sewage, plastic and industrial wastes. Not only the quality but the quantity of available water is also declining. Irrigation allows cultivation of great areas which would otherwise be barren. Nearly three million square kilometres of land are irrigated, an area almost the size of India. Yet, although the irrigated area is increasing, only about one-third of the irrigation water reaches plants. The result is a net loss of ground water available for cultivation. In many regions, in addition, evaporation raises salt concentrations and makes soil unsuitable for crops. Today, for this and other reasons, much cultivated land is being lost.

A report from the World Bank suggests that the wars of the twenty-first century will be fought over water. The Middle East, much of which is desert, is the leading example of the problems of water supply which, like those concerning locusts, can be solved only by cooperation between nations. Despite wars (and the resulting pollution), attempts at agreement are being made. Northern Africa, parts of India and Pakistan, northern China and much of Central and South America face similar dangers and need the same cooperation among governments.

A Struggle for Survival

The actual future will therefore depend not only on the climate but also on the climate of opinion. A leading feature of the social history of the nineteenth and twentieth centuries is what people have done to improve or to preserve the health and environments of human populations. The story is one of struggle. Proposals for remedy, however urgent, have been forcefully resisted.

Today, in many countries, we are accustomed to a high standard of hygiene: we do not defecate in the streets or even in the fields; we do not throw our slops out of the window or our kitchen refuse over the wall. As a result, we no longer die young of waterborne diseases such as cholera and typhoid fever; nor is gastroenteritis a leading killer in infancy.

The 'sanitary idea' was first accepted, in the countries of western Europe, only in the second half of the nineteenth century and later still in the United States. A specialist in public health, A.J. McMichael, suggests that 'even more than parliamentary democracy, railway systems and cricket, England's greatest contribution to modern civilisation may have been civic hygiene'. Yet, during decades of conflict, its supporters had to overcome great resistance, both from bureaucracy and also from those who stood to lose by it. The great English reformer, Edwin Chadwick (1800–1890), has been described as fighting a bitter and often futile battle for clean water. Members of the government in England were shareholders in water companies. They therefore opposed laws to regulate water supplies and sewerage systems. This was a time of outbreaks of waterborne cholera in London. A reliable supply of uninfected water, which many readers today can take for granted, has been generally available, even in the most advanced countries, for less than a century. In much of the world, it is not available yet.

A more recent example concerns the use of insecticides, of which the best known is DDT. These substances, as we know, had at first great success against pests; but they soon proved to be dangerous to other life: they have therefore been called biocides. Human beings, domestic animals and wild animals and plants have all been killed or harmed by them. Large predators have been especially affected. DDT residues have been found in penguins and seals in the Antarctic, thousands of kilometres from the sources of contamination.

In 1962, an American science writer, Rachel Carson, published a book, *Silent Spring*, on these dangers. She was at once energetically denounced as a scaremonger and called 'a fanatic defender of the cult of the balance of nature'. Although, in rich countries, such as those of western Europe and North America, legislation now regulates their use, these poisons continue to be manufactured at an increasing rate.

As the twentieth century ends, warnings on the effects of pollution, on the enhanced greenhouse effect and on the destruction of forests, soils and waters are meeting enhanced hostility. In 1990, from the American centre of trade and money, the *Wall Street Journal* published a leading article on environmentalism: it complained about 'scientific faddism' and criticised predictions of global warming. Other writers have written of 'green hysteria'. An Australian article on this theme states that it would be foolhardy to succumb to the notion that governments must intervene to bring economic development to a halt or even to curtail it. The free enterprise and market economy promoted by this journal evidently requires total freedom for private production, without regard to any adverse consequences for human beings now or in the future. Late in 1991, in the United States, the Climate Council, a lobbying organisation of the United States energy industries, even protested against the efforts of climatologists to estimate carbon dioxide levels in the year 2010. In 1995, representatives of two oil producing countries, Saudi Arabia and Kuwait, tried to dilute a report on the greenhouse effect for fear of its adverse effects on profits from oil.

The formidable skills of experts on public relations have now been mobilised to discredit the environmentalists. An American organisation, Wise Use, with collaborators in many countries, urges opening national parks, forests and wilderness for mining and for oil and gas. Its declared aim is destruction of the environmental movement. It therefore labels peaceful conservationists as terrorists and fanatics, with a secret program for a totalitarian or communist world government.

One consequence is the continued destruction of natural resources. Many can be restored only after centuries of effort. The entrepreneurs responsible for converting South American forests to temporary pasture planned their operations for a few years. They made large profits but left wilderness (chapter 8). Other instances mentioned in this book include prolonged overfishing of cod, herring and other species and the near extinction of whales and other large mammals. At first nobody knew what was happening to these populations. Today, we know a great deal: governments and industry may no longer justify neglect of global ecology by pleading ignorance.

Nor may economists. Until recently, modern economists have usually taken it for granted that the capital needed for food production—water, soil, forests and fisheries—will be available however much is consumed. Similarly, they have 'externalised' the costs of pollution and of injury to health, that is, they have ignored them. Their calculations have applied, therefore, only to the short term.

These limitations are increasingly recognised. The conventional index of a nation's economic growth is the gross domestic product (GDP), which measures only the rate at which money is exchanged. A group of American economists, called Redefining Progress, point out that the individual who contributes most to the GDP in the United States is a rich terminal cancer patient (gigantic medical expenses) who is involved in an expensive divorce (exorbitant legal fees). The GDP therefore not only ignores but also hides much of what is going on in the real world. Some economists and others talk of sustainable growth; but, if the deficit of capital resources is allowed for, the growth becomes a loss.

As the century ends, human numbers continue to rise and consumption of food and other resources with them, while food production stagnates. A crucial question is therefore how far people are willing to look ahead. In the 1990s, small countries and clans, armed for profit by more powerful nations, are still fighting for soil which, as the combatants kill each other, is being polluted or ruined. If this mass suicide continues, it will leave only starving remnants of humanity scratching a living in wastelands.

All the inhabited world is under threat. At worst, by the year 2035 central Europe and the American Midwest will be deserts. Fires will then destroy much of the world's few surviving deciduous forests. The Nile Delta, Bangladesh, parts of Vietnam and the Netherlands, all at present densely populated, will be under water. People will emigrate in overwhelming numbers: the population of Canada could go from 20 to 200 million. In crowded northern Europe malaria will return with the higher temperatures: new, virulent strains of malarial parasites will be carried by mosquitos resistant to insecticides. Other infectious diseases, such as plague, will be out of control. Famine will be widespread. Political organisation will collapse. Already, in the 1990s, we have floods in China, India, Europe and the United States; famine in Africa; forest fires in the United States and Australia; and social upheaval in the Middle East, the Balkans and elsewhere. In 1997, as the proofs of this book were being marked, enormous areas of South East Asia, especially Indonesia, disappeared in toxic fumes from burning forests; the pall spread even to the north of Australia.

BICYCLES, COOKING STOVES AND COAL EQUIVALENTS

That 'worst case scenario' does not represent a foregone conclusion. All major governments now agree that the biosphere is deteriorating daily and are beginning to accept the need for global regulation. In the 1990s, delegates from over seventy nations announced a program of action against greenhouse gases. In 1991, new satellites were launched to watch changes, especially harmful ones, in the biosphere. In 1992, all major countries and most of the minor ones were represented at an Earth Summit. In 1993, an International Convention on Biological Diversity came into effect: governments are required to prevent losses of irreplaceable ecosystems, especially certain kinds of forest; uncontrolled logging is to be replaced by regulated planting and harvesting of timber. Brazil, with its enormous devastated areas, is now turning to a massive program of reforestation. In a drastic change of policy, the program is being financed by the World Bank. New planting is also aided by a tree bank which collects, preserves and grows the seeds of threatened species.

For effective action of this kind, managers of states need to take decisions while faced with many uncertainties. The predictions of economists, on which politicians partly depend, are even less accurate than those of climatologists. So are the predictions of others. In a survey by a London journal, *The Economist*, covering ten years from 1984, people of several occupations were asked to foretell changes in the world's economy. English dustmen (garbage collectors) tied with company chairmen; finance ministers from four countries scored lowest.

If action is delayed until scientists are unanimous about the exact consequences of the enhanced greenhouse effect, about the exact outcome of destroying forests and rangelands, about the exact scale of the losses of food from fisheries, in short, about the exact form of the future, then massive disaster for humanity will be ensured. Even useful calculations are always based on incomplete information. The incompleteness is partly of our own making: decisions taken now influence the future.

Some crucial decisions concern the use of energy. Energy consumption is conveniently measured in CE (coal equivalents). In 1991, reports from the United Nations estimated India, the largest 'developing' country, to consume 250 CE annually for each head of population. At the other extreme, the figure for North America was 11 000 CE. The figures of energy consumption reflect the differences in total wealth between rich and poor countries. A more general index is the gross national product, or GNP, which resembles the GDP mentioned above. Divided by the population of a country, it gives a measure of the average level of monetary riches. According to the UN, in the 1980s the figure for the twenty richest countries was US$1300 and was rising steeply. For the 33 poorest countries the figure was US$270 and was hardly rising.

In 1974, the United Nations had adopted a Declaration on a New International Economic Order: the stated objective was to eliminate the gap between rich and poor countries by accelerating economic development. But, as the UN has now shown, if the poor countries attempt a level of consumption similar to that of the richest countries and people, they will only make the impending environmental calamities more severe. For the whole of the present world's population to consume existing resources at the highest present rate is indeed an impossibility. The implications of this fact have still to be generally recognised. They include the need for drastic reductions in expenditure by the wealthy in the richest countries, especially on petroleum; and, more important, they require a fundamental shift in the assumptions commonly made about what may be consumed or wasted.

Questions concerning the biosphere arise in all societies and regimes, regardless of their social or political presumptions or 'ideologies'. Even if all governments, red, blue or black, turned green overnight, the obliteration of species and habitats would continue at least for decades. On the problems of pollution, governments are unlikely to use Edward I of England (1239–1307) as a model: he advocated torture and execution for people who burnt coal. But extensive interference with the freedom of both industry and private persons to poison the air and the waters is needed. At present, anyone who can afford it can fly or go by car almost anywhere he or she chooses. This licence will have to be surrendered. The frenzied efforts to encourage tourism and to sell more automobiles will have to stop. Ninety-five million bicycles were made in 1990—a record number; many more will be needed.

The governments of poor countries may decide to build more nuclear power stations and so reduce the output of carbon dioxide. If so, they will increase hazards already serious and well known: disastrous accidents have occurred in the United States and the Ukraine; large numbers of workers in mines, processing plants and power stations have been dangerously irradiated; and, still worse, no satisfactory method has been devised for disposing of radioactive wastes.

Other, nonpolluting sources of energy are available, all of which could be used much more than they are now. They include the sun (for heating buildings and water), wind (for windmills which generate electricity) and water, including tides, (for hydroelectric systems). Wind power is among the most promising: the winds of Britain alone could provide the energy needed by much of Europe; similarly, North and South Dakota could supply the United States.

Yet another proposal is the use of geothermal energy. Most of our planet is very hot. Not far below the surface are rocks at 200°C. Heat from rocks is always there regardless of the weather. Water can be pumped through them and the heat used to drive power stations without polluting the atmosphere. Some large buildings are already supplied in this way.

Meanwhile, knowledge of how to restore soil fertility and to grow more crops and trees has, in a few decades, greatly advanced. In the 1940s, India still had famines with a high death rate; but, well before the end of the century, Indian agriculture was regularly producing a small surplus. The green revolution which made this possible depended on three things: new varieties of crop plants, especially rice, were produced; the use of artificial fertilisers was matched to the crops; and irrigation was made more efficient.

This was a notable achievement but it left many problems unsolved. If rice is more widely or more intensively cultivated than before, it uses up much water; ground water levels then fall, with destructive consequences after a few years. Irrigation, as we know, can also lead to accumulation of salt in the soil and a resulting decline in soil fertility. To prevent these ill effects, central plans are needed. At present, rice and wheat are often sown where it would be better to encourage cultivation of local crops such as manioc (cassava, *Manihot esculenta*) or millet (*Panicum miliaceum*).

The conventional picture of successful, modern farming is of rolling fields each with a single species of food plant (monoculture). This is a legacy of a stable European agriculture which developed over many centuries. In the twentieth century, the highly mechanised western agronomy has been rapidly transferred to other regions at the cost of much deterioration of soils. Monoculture leads also to severe losses of biodiversity: a few species and varieties are grown at the expense of a large number of alternatives. It therefore increases the dangers from pests and diseases.

Modern farming includes great herds of cattle bred for beef. To produce 5 kilograms of beef protein requires nearly 80 kilograms of plant protein. The use of diminishing fertile soil to yield meat is therefore grossly extravagant: it is better to grow plants which can themselves be eaten by people. Vegetarian meals will soon be unavoidable by many at present accustomed to steaks and roast joints.

Farmers in many regions have already adopted polyculture (or permaculture) of plants which directly yield human food: rice may be accompanied by legumes, such as lentils and beans, and other species, such as sunflowers, which produce edible oils. Many plants grow best neighboured by other species. In southern Africa, permaculture is increasingly used to provide stable and productive farming.

The importance of small things has now also dawned on the experts in a big way: cooking stoves, for instance. The food of villagers is usually cooked by burning dung, wood, plant residues or coal in small stoves. Most of the heat is dissipated and much injurious smoke fills the dwelling. More efficient stoves have recently been designed: an important feature is a ceramic lining to retain the heat. The users of the stoves were at first ignored but, once they were consulted about details of shape and size, use of the new stoves spread widely. About half the world's population depends on such equipment. For them, attention to domestic matters can lead both to large savings of fuel and also to improved health.

At the same time, in almost all countries, including the poorest, wooded country offers many neglected opportunities. An English biochemist, N.W. Pirie, has shown how inedible leaves can yield a nutritious protein extract equivalent to cheese. Admittedly it needs spicing up if it is to be enjoyed by human beings. It can, however, also be used in fodder for animals. Several hundred species have been treated in this way, especially in the tropics. Manufacture of leaf protein is simple and can be done in villages.

Here are some further examples of how planning can be combined with attention to local needs. In India, for over a century, successive governments have tried to manage forests by edicts issued from above. During that period, the population has more than doubled. As a result, in enormous areas, forests have been stripped and the soil eroded. Late in the twentieth century, India was losing 1.5 million hectares of forest yearly, an area similar to that of the State of Connecticut in the United States.

Most Indians live in small villages. In many places the villagers wish to draw on the forests for fuel and for building; they need its products also for farming equipment and thatch and for many foods and herbs. The government foresters, for their part, wish to preserve the forests, but often for the single purpose of providing poles. The result has been a standoff, sometimes violent, which did nothing for either side.

But, by the 1980s, policies were changing. In West Bengal large areas have come under the joint management of villagers and forestry officials. The villagers are no longer kept out: they can feel that the land belongs to them, as it did long ago. In a few years, sal trees (*Shorea robusta*), an important source of timber, have regenerated from stumps and are surrounded by productive undergrowth. More trees are being planted. The villagers can again make use of forest products; and, when timber is sold, they receive some of the profits. Volunteers protect the land from poachers.

One factor in the success of West Bengal is a State government which favours land reform and programs designed to help the poor. In other parts of India, threatened forests have received another, more portentous kind of protection. The Chipko movement was begun by mountain women in the Himalaya but has spread to other regions, including some far away in the south of the country. The women have sought to preserve trees from destruction by all means, including putting their bodies in the way of loggers. Some women have died; but the movement is growing.

Each region of the world has its own problems. The small Central American country of Costa Rica is a leader in one kind of reforestation. The dry forests along the Pacific coast had an immense wealth of plant and animal species largely destroyed by unplanned farming. Ecological research has made restoration possible. The seeds of some essential tree species were originally spread by large mammals, now extinct. Part of the remedy is to allow free movement of cattle and horses between pastures and forests. As in India, both the support of the farmers and special financial measures are crucial.

Further north are the vast temperate rainforests of the North American West Coast, from California to Alaska. Despite anarchic logging, some superb old growth remains, including cedars which were seedlings at about the time of the Norman conquest of England in 1066. The forests are part of an ecosystem in which the many rivers support a great wealth of edible species. The most important are salmon and oysters, but clams, crabs and others are also present. Among the fish is a dwindling population of the magnificent sturgeon (*Acipenser sturio*) which in English waters is, by a fourteenth century Act, the property of the monarch.

The fisheries depend on the continued existence of highly productive forests. As the forests have been cut down, pollution has risen and productivity has steeply declined. The human population has been impoverished and unemployment has risen.

But much research has been done (more is, of course, needed). Regional organisations, supported by banks which favour small 'green' enterprises, are now beginning to make rational use of the resources. Jobs are being created.

These histories are part of a worldwide movement: its objective is rapid action, on both large and small scale, to combine conservation of nature with human survival.

The 1990s have also seen a beginning, by the most heavily armed nations, on disarmament. If this continues, massive additional resources will become available for constructive ends. The transformation required is, however, not only a matter of beating swords into plowshares. The need to restrain human numbers has still to be adequately faced. An animal population may destroy its sources of food or shelter, or it may incur high rates of mortality or infertility as a result of crowding: it then declines or becomes extinct. We differ from other species in our use of reason and in the ability to plan ahead: we can take decisions not only on next year's crops but also on policies for the next decades or century. A radical example of planning comes from China, where the government requires each married couple to have only one child. Such a strategy, even if it works, is unlikely to be copied in other countries. It does, however, reflect an awareness of present dangers. So does the beginning acceptance elsewhere of a new obligation on parents: not to have more than two children.

Whatever schemes are adopted, extensive and rapid social changes must be expected, as great as the collapse of civilisations based on slavery or the replacement of feudal order by production for the market. Pestilence and erosion of the biosphere need not be unmixed disasters: they can lead to a moral revival in which fixation on immediate private gain is replaced by communal action.

So far, we have effective general agreement only on special requirements. Many international and national regulations are based on biological principles: among the most important are those designed to prevent infectious diseases such as cholera and plague. As well, farmers in some countries are obliged to grow certain crops and not others. It is also becoming usual to demand an 'environmental impact statement' before new projects, such as an oil rig, an airport or a very fast train, are undertaken. While this book was being written, a large factory was built in the south of Germany to satisfy all the demands of 'ecological puritanism'. Thousands of trees and shrubs were planted in the factory grounds; a road tunnel allows preservation of some ancient trees; pollution is avoided by the use of two, separate sewerage systems. The buildings are kept warm by surplus heat from the machines. Workers move about the factory on bicycles.

Survival in the twenty-first century will depend on an immense expansion of such measures. For them to succeed, large numbers of biologists and other people with special skills will be needed. Many employed in 'defense', that is, war, will perhaps turn to the arts of peace.

Specialists, however, can usually answer only special questions. Success in killing insect pests has led to medical and economic hazards which cannot be dealt with by chemists on their own. Similarly, although an expert may state with confidence how a wilderness can be converted back to woodland, the attempt to do so inevitably meets an array of social obstacles: what are the needs and preferences of the people who inhabit the wilderness? We need, as well as specialists, generalists trained to cope with such situations.

Shortage of highly trained people is one aspect of the world education crisis. Another is the existence of hundreds of millions of people who cannot read a newspaper or write a letter. In the United States, the richest nation, at least forty per cent of adults are functionally illiterate. This has been widely known, and frequently confirmed, since Jonathan Kozol published his *Illiterate America* in 1985; but the remedy is not in sight. Such people are severely handicapped in understanding what is going on in the world; they cannot take a full part in the political life of a modern democracy.

The tasks for humanity in the twenty-first century demand prolonged and informed debate on the biological threats and opportunities we face and the policies we should adopt. Debate should lead quickly to action. In 1966, an American ecologist, Aldo Leopold, proposed an 'ethic of conservation' for all humanity. The idea was not new: the principle of preserving one's living heritage for one's children and their descendants is prominent in the attitudes of those who live by cultivating the soil, and of tribal peoples. It is, however, often lacking among modern city dwellers and those who engage in large scale agribusiness. In much of the world, a dominant principle today is rapid growth in immediate monetary profit. This moral bankruptcy is combining with the rise in human numbers to destroy the biosphere.

At the opposite extreme is the concern, felt by some people, about the existence of environments and species which they will never see, except in pictures. Many readers would be upset if blue whales, rhinos, elephants and others were exterminated; as were, in living memory, the quagga and the thylacine. The concern goes with willingness to contribute to their survival. Some economists therefore now talk of the 'existence value' of threatened species and ecosystems: they recognise that people do not always make decisions solely on the basis of cost effectiveness or private convenience.

Finding out about the millions of species, how they live and how to live with them, could keep our descendants happily occupied for centuries, even millennia. We can imagine, while in buoyant mood, a future global community of which a central feature is the study and enjoyment of nature. Provided that it does not become a form of escapism, at least this fantasy may help to remind us of the scale of the tasks ahead.

A serious hazard is paralysis due to dismay at the size and variety of present problems. Most of the preceding chapters give examples; but they also indicate ways in which the problems are being solved. Although central planning on a large scale is always needed, success also requires local action: any adult in a democracy can contribute. Our accumulated decisions, large and small, about the living world, taken in the next few decades, will influence the kinds of lives our successors lead for centuries to come.

Appendix I

A Classification of Organisms with Miscellaneous Annotations

ALL THE MAIN GROUPS (taxa) are listed, plus some oddities. Systems vary: this one is chosen for convenience. Ordinary names are used wherever practicable. Because systematic lists of names and formal descriptions are not much fun to read, some unsystematic notes are added.

Prokaryotes (Monera)

Microscopic organisms without cell nuclei. Nearly always single cells. They give us a notion of how life began and today are matters of life and death for humanity.

Bacteria

Bacteria usually have no chlorophyll. Two vast groups, both chemically very diverse. Most of the more familiar EUBACTERIA are decomposers, especially in soil or water; but many are parasitic and cause disease. The ARCHAEBACTERIA include forms which live in extreme conditions, such as salt lakes and hot springs; others are present in the stomachs of ruminants and in compost and produce much methane—a 'greenhouse gas'.

Blue-greens (Cyanobacteria)

Blue-greens have chlorophyll. From way back, probably principal creators of our oxygenated atmosphere. Can live on bare rock and help to begin soil formation.

All other organisms are eukaryotes: that is, most of their DNA is in a nucleus.

PROTISTA

Microscopic organisms, usually single-celled. Another enormous, heterogeneous collection. May be crudely divided as follows.

Protophyta

These are single-celled, green organisms, often classified with the plants. Some have flagella, and so are mobile like animals. In the sea, one group, the diatoms, are the most important photosynthesisers. On land some of the Protophyta make the green scums which develop in stagnant waters.

Protozoa

Protozoans are single-celled organisms often classified as animals. They include *Amoeba* and *Paramecium* of the textbooks, also the malarial parasites.

Slime fungi

These are completely weird and have been claimed both by zoologists, as Mycetozoa, and by botanists, who have put them with the fungi as Myxomycetes. The best studied, *Dictyostelium*, lives among leaves on forest floors. An airborne spore touches down, becomes a small ameba, feeds on bacteria and yeasts and divides repeatedly. When the amebae run out of food, they collect in a 'slug' which moves toward dim light and leaves a trail of slime. After about 30 hours, it grows a vertical fruiting body which contains thousands of spores and resembles that of a fungus but has a wall of cellulose, a substance typical of plants. If a slug is allowed to flow through silk gauze, with pores of less than 0.1 mm in diameter, it forms again on the other side and flows on. This has been likened to a ghost going through a brick wall.

PLANTS

Many-celled green organisms most of which have firm cell walls and are capable of photosynthesis. Complex responses are made to stimuli, but without organs of active movement.

Algae (Thallophyta)

Algae are the seaweeds and their relatives. The reds are the most edible but even the brown kelps, prominent on temperate rocky shores at low water, can be ingested at a pinch: try boiling them in milk. Of course, some algae break the rules. Among the species living on coral reefs are green plants of the genus, *Caulerpa*, of which one is eaten in salads in the Philippines. The leaflike structures grow up from what look like stems and roots to make a plant up to a metre long. The tissues have enormous numbers of nuclei but *no cell walls*. The 'leaves' are supported by fibrelike microtubules. In biology, always something unexpected.

Mosses (Bryophyta)

Mosses are land plants but need wet conditions. They have an obvious alternation of haploid (sexual) and diploid (asexual) generations; no roots or strengthening woody tissues. Relics of the remote past.

Ferns (Pteridophyta)

Ferns also have alternation of generations, but the gametophyte is small and obscure. They have roots and woody tissues but no flowers. Some form substantial trees. The largest are extinct: coal consists largely of fossil ferns.

Conifers (Gymnospermae)

Fir trees and relatives include most of the evergreen trees. The gametophyte is a mere vestige. No flowers. Extinct forms are prominent in rocks as early as the Devonian period. Provide us with about 75 per cent of our timber; also secrete resins which give us turpentine, oils and tar.

Flowering plants (Angiospermae)

Angiosperms have flowers which produce seeds enclosed in an ovary. The gametophyte is again a vestige. Reproduction requires the transfer of pollen, usually from one flower to another. More than 235 000 species, from duckweeds (Lemnaceae), only 1.0 mm long, to the mountain ash (*Eucalyptus regnans*) of Tasmania, height more than 100 m, and the redwoods of California of similar height and even greater girth. Redwoods of the species, *Sequoiadendron gigas*, are (except perhaps certain fungi) the most massive organisms now living. The DICOTYLEDONAE include most of the flowering plants. The embryos which grow from seeds each have two cotyledons (seed leaves). The veins of the leaves are a network. The parts of the flowers (petals and others) are in fives, fours or twos. The MONOCOTYLEDONAE include the grasses, lilies and orchids. The embryo has only one cotyledon. The leaves have parallel veins. The flower parts are in sixes or threes.

Butterflies and moths (Lepidoptera) and bees (Apidae) are symbiotic with flowering plants. Some Lepidoptera have enormously long proboscies which, when expanded, exactly match the spurs of particular flowers. Part of a bee's mouth (the labium) is a tube, sometimes very long, which sucks up nectar which, in the nest, becomes honey. The 'hairs' on the body are feathery and collect pollen which is transferred to pollen baskets on the legs or body and carried home for food; but some pollinates other flowers.

Most, perhaps all, flowering plants and conifers are also aided underground by fungi. This is mycorrhiza ('fungus root'). 'Mycorrhiza' sounds like the name of an organism; 'fungus root' may suggest a disease. In fact, however, this is a symbiotic relationship. The 7000 species of orchids are especially dependent: after germination of the seed, they can grow only in the presence of particular species of fungi.

Charles Darwin once arranged for a trombone to be played to a climbing plant. The plant took no notice. A century later, many people were led to believe that plants have a 'secret life' and react to human feelings and to music (especially that of J.S. Bach) and can predict the weather; but a number of careful workers failed to confirm the original report.

Amazing to relate, however, some plants have a much higher temperature than their surroundings. A few are, like birds and mammals, homeothermous (one can hardly say warmblooded). The flowers of one, the sacred lotus (*Nelumbo nucifera*) maintain a temperature of up to 37°C in surrounding temperatures down to 10°C. The fevers of some flowers release volatile substances which attract pollinating insects; and, in cold environments, some flowers attract insects simply by their warmth.

Plants also respond, more familiarly, by changes in growth, to external influences, especially light, gravity and contact: stems grow toward light (phototropism); roots toward the earth's centre (geotropism); climbers' stems curl round objects in response to contact. The growth rates of the main growing regions (meristems) of a young plant, at the tips of the stems and roots, depend on growth regulators, or hormones, notably auxins. The growing shoot of a pineapple plant (*Ananas comosus*) has a concentration of auxin of about 6 micrograms to every kilogram of tissue—about the weight of a needle in a haystack weighing 22 tonnes. Yet auxins have been put to use. Fruits normally grow only after a flower has been pollinated; they then contain seeds. Auxins can stimulate fruit formation without seeds, as in many of the grapes we now eat. Other hormones, the gibberellins, stimulate breakdown of starch to sugar during germination of wheat, barley and oats. They can also stimulate the flowering of crops such as cabbages and carrots and therefore the production of seeds.

Plant eaters and parasites are put off by pungent or poisonous substances in the plants' cells. The leaves of Australian eucalypts are exceptionally toxic. Many green plants also produce antibiotics (phytoalexins) which prevent the spread of bacteria and moulds. Methods are now being developed to stimulate crop plants to produce these substances *before* they become infected. Plants also synthesise insecticides. Pyrethrum (used in our houses) comes from a species of *Chrysanthemum*. The larvae of the gypsy moth (*Lymantria dispar*) sometimes attack oak trees (*Quercus*) in vast numbers and destroy their leaves. The trees stunt the larvae by growing tougher leaves with extra poisons (such as tannins), but less water. Willows (*Salix*) and alders (*Alnus*) too are attacked by defoliating caterpillars. The trees then not only make their leaves unpalatable but also secrete a volatile substance which evaporates and is carried to other trees. The latter, even if undamaged, respond as if they too had been attacked.

So plants have a secret life. It is exciting enough, but it is not full of human emotions: it is plant life.

Fungi (Eumycophyta)

Resemble plants in being without organs of locomotion and in having firm cell walls, but are like animals in depending on complex organic substances for nourishment. The ASCOMYCETES include the yeasts, mildews and cheese moulds. The PHYCOMYCETES are the bread and leaf moulds. The BASIDIOMYCETES are the mushrooms, including honey fungi, and rusts.

The honey fungi enter live trees, invade their roots, grow into the trunk and destroy the conductive tissues. The trees die but the fungi live on the dead wood. A single *Armillaria bulbosa*, in Michigan, USA, occupied 15 ha of woodland, is believed to have been 1500 years old and may have weighed more than 100 tonnes (much the same as a blue whale).

The mushrooms and toadstools produce spores in large fruiting bodies. One species, *Agaricus bisporus*, is cultivated in at least seventy countries. The common edible mushroom of temperate lands is a similar species, *A. campestris*. The death cap (*Amanita phalloides*) causes most of the accidental poisonings due to fungi. Eating even part of a fruiting body can be fatal, but a tiny fragment can be consumed to give a few hours of ecstasy. The Kamchatkan nomads drug themselves by drinking the urine of people who have eaten *Amanita* and survived. Not recommended.

Animals

Depend for food on taking in complex organic substances. Usually mobile at some stage. No firm cell walls and no chlorophyll.

Sponges (Porifera)

The sponges are very odd: not mobile, have no organs and live by straining food from water. A single sponge looks like a very large colony of flagellate protozoans supported by a skeleton of calcareous spicules. Present in all seas (and some fresh waters) which offer surfaces for settling. The cells of a sponge can be separated by forcing them through a fine sieve; they then reassemble in groups, some of which develop into new, complete individuals. A few sponges put out filaments covered with hooks and snare prey such as small crustaceans; cells move out, digest the victim and return to the body of the sponge. Of the 5000 species, the bath sponge (*Euspongia*) is a Mediterranean form in steep decline owing to pollution. Sponges are now being studied intensively because some contain antibacterial and antiviral substances.

Coelenterata (Cnidaria)

Coelenterates have a radially symmetrical body with only two distinct layers of cells (ectoderm and endoderm). The ectoderm has stinging cells. The nervous system is a network. The simplest form, exemplified by *Hydra*, is a single sac with an opening surrounded by tentacles; but many species are colonial. Corals, the most prominent colonials, secrete a mineral platform and can form long reefs which fringe shores; in deep waters they can make gigantic barrier reefs; and in some tropical waters they create atolls (coral islands) many of which support trees. The largest, in the Maldives, is 140 km long and 32 wide. A large coral reef shelters hundreds of species of fish, molluscs, crustaceans, worms, starfish, algae and others.

These ecosystems have long supported fisheries in Pacific and Caribbean waters, but recently they have been almost annihilated by two enemies. (i) The crown-of-thorns starfish (*Acanthaster planci*) has a dense covering of hard spikes and may reach a diameter of 60 cm. Since about 1960, in many parts of the Pacific, it has undergone an unexplained population explosion and killed more than 90 per cent of the corals. The denuded, dead coral is soon covered with algae but the fish and many other animals depart. (ii) In the 1980s, over great areas, at a time of exceptionally high water temperatures, the corals have turned white and died. Is this a consequence of global warming (the greenhouse effect)?

All other animals have three cellular layers and are bilaterally symmetrical at some stage.

Flatworms (Platyhelminthes)

The flatworms have vertically (dorsoventrally) flattened bodies and a central nervous system. They have a mouth but no anus, and no circulatory or respiratory systems. They include many parasitic forms (flukes and tapeworms).

The FLUKES (TREMATODA) live, as a rule, in the liver or blood of other animals. The three species of *Schistosoma* (also known as *Bilharzia*) infect millions in the tropics of Asia, Africa and South

America. Like the malarial parasites, they cause chronic weakness, debilitate large populations and kill many children. Eggs released in human urine or feces are ingested by freshwater snails and develop into larvae; these escape into the water and can enter a human being through the skin or in drinking water. Some eggs are eaten also by fish which may then be eaten by people. Hygiene is the main remedy.

The TAPEWORMS (CESTODA) are more imposing than the flukes but less important. The beef tapeworm may have over 1000 segments and be more than 6 m long—an unwelcome occupant of the human intestines. Cooking and inspecting meat prevent infection.

Roundworms (Nematoda)

Nematodes form an enormous, ubiquitous but unobtrusive group of mostly very small worms. The body is cylindrical with pointed ends and is covered with a tough cuticle. A mouth is near one end and an anus near the other. Most live in soil or water, but many are important parasites in plants or animals, including ourselves.

The large roundworm, *Ascaris lumbricoides*, up to 30 cm long and 4 mm in diameter, lives in the human gut and absorbs predigested food. The female lays vast numbers of eggs which pass out in the feces and, where hygiene is poor, enter the soil and the water of streams and rivers. The population of China has been estimated to produce 18 000 tonnes of *Ascaris* eggs annually. Reinfection is by swallowing eggs. These worms, like other such parasites, cause debility among millions and kill many children.

The same applies to the hookworms (*Ankylostoma* and *Necator*) which attach themselves to the gut wall, feed off blood which leaks from injured blood vessels and cause severe weakness due to anemia. In warm, moist soil the eggs, voided with the feces, develop into small worms which readily enter the skin of bare feet, where they cause intense itching. Rural populations debilitated in this way include those of southern States of the USA. Treatment with drugs can be effective. Prevention requires improved hygiene and wearing shoes.

Some nematodes have complex life cycles. The adults of *Wuchereria bancrofti* grow up to 100 mm long, are only 0.1 mm wide, live in human lymph nodes, block the flow of lymph and so produce the hideous deformities of elephantiasis. This is yet another 'tropical' disease present in tens of millions of people. The females release larvae (microfilariae) which enter small blood vessels in the skin where they are picked up by blood-sucking insects. In the insects, the larvae develop to a stage at which they can reinfect a human being. The disease can be treated by drugs.

Mollusca

Molluscs typically have a very hard shell but some are slugs without shells. They are not segmented. Most move by means of a single 'foot'.

The GASTROPODA include the snails, slugs and whelks. The BIVALVIA are the clams, mussels, oysters and others. The CEPHALOPODA are mostly highly mobile and include the squids and octopus.

The largest bivalves are the giant clams (*Tridacna*) some of which, embedded in the coral of the Great Barrier Reef off the east coast of Australia, reach a length of 1.0 m. Plans are being made to farm them for food, like oysters (*Ostrea*).

The largest cephalopods are squids, which have ten tentacles equipped with suckers, are ocean dwellers and can move swiftly by a form of jet propulsion. The mouth has a pair of beak-like teeth. The eyes are the largest in the animal kingdom. The giant squid, *Architeuthis* (or Kraken), can digest small sharks and dolphins. The largest recorded was about 19 m long with its tentacles; the body made about half this. Stories about it often sound like fiction and no doubt some are; but a giant squid can certainly sink a small boat and even perhaps ingest its crew.

Cycliophora

In biology always something new. In December 1995, two Danish zoologists reported a new phylum with one species, *Symbion pandora*, less than 1.0 mm long. It has cilia round its mouth and drives small particles into a U-shaped gut. The anus is near the mouth. It reproduces both sexually (the male is described as a dwarf!) and asexually by budding. It was found on the mouthparts of a lobster, *Nephrops norvegicus*, regularly eaten by gourmets. The adult *Symbion* sticks to the lobster by a stalk. An enthusiastic zoologist has urged waiters in seafood restaurants to have microscopes and textbooks of invertebrate zoology available for their clients.

A recent textbook lists fourteen phyla, most of them obscure, under the heading, 'Worms'. From them, here is one.

Pogonophora

Pogonophora live mostly in deep waters heated by volcanic action, are long, thin and gutless, have a tuft of tentacles at one end and derive nourishment from symbiotic microorganisms. They have only recently been studied in detail. Rather bizarre.

Ringworms (Annelida)

These familiar worms are segmented, have a distinct head and a closed circulatory system, a well developed ventral nerve cord with a ganglion in each segment and a brain in front. The gut runs the whole length of the animal.

The BRISTLE WORMS (POLYCHAETA) live in the sea and have paddle-like limbs (parapodia) with bristles. Many live in tubes. Lugworms (*Arenicola*) are much used as bait. The EARTHWORMS

(OLIGOCHAETA) live mainly in soil or fresh water and have bristles but no parapodia. Some eat their way through soil and contribute greatly to its fertility. The LEECHES (HIRUDINEA) have no parapodia or bristles, only suckers. Many suck blood and were once used by physicians (also called leeches) who literally bled their patients. They are now coming back into use to dissipate accumulations of blood in damaged tissues.

Velvet worms (Onychophora)

The velvet worms make a very small group, with a long fossil history, traditionally believed to be intermediate between the Annelids and the Arthropods. Each of their many segments has a pair of clawed legs. The head has a pair of antennae. Respiration is by tracheae.

Arthropoda

A vast group of segmented animals with a hard external skeleton of chitin and jointed legs; the body is divided into a head, a thorax and an abdomen. The CRUSTACEANS have a head with two pairs of antennae, respire by means of gills and live mostly in the sea. Each abdominal segment usually has a pair of limbs used in locomotion. They include water 'fleas', deliciously edible crabs and lobsters, barnacles and terrestrial woodlice (sowbugs).

The INSECTS have a head with four pairs of appendages; the thorax has three pairs of legs and usually two pairs of wings. The abdomen has no walking legs. Respiration is by tracheae. The number of insect species is greater than that of all the rest of the animal kingdom.

The ARACHNIDS have no antennae: the first pair of appendages are pincers, the second are jaws and the last four are usually for walking. Respiration may be by gills or by tracheae. Most live on land and are carnivorous—spiders, scorpions, mites, ticks and others.

A spider has a combined head and thorax joined to an abdomen by a narrow waist; it moves on eight legs. The head usually has eight eyes and a mouth equipped with poison fangs. Prey are injected with poison and digestive secretions and are sucked dry. Silk glands and spinnerets weave webs and other structures. The elastic silk of an orb weaver, apparently flimsy but of astonishing strength for its weight, has a breaking strain much superior to that of steel. Though secreted as a watery mixture it is insoluble (hence not dissolved by rain). Its mechanical properties depend on alternating orderly and disorderly arrangements of large protein molecules. Efforts are now being made to manufacture it: by engineering their genes, bacteria have been made to produce granules of spider silk. Many applications are expected, for instance in surgery.

A web-spinner, the black widow (*Latrodectus mactans*), widespread in warm climates, is one of the few dangerous species. The closely related redback spider of Australia is equally toxic. A bite may not be noticed but several painful days are likely to follow and then (usually) recovery. An antidote (antivenin) is available. Among the active hunters, the Carolina wolf spider, *Lycosa carolinensis*, of North America, can reach 35 mm. The bite of the European tarantula, *L. tarentula*, has been supposed, probably incorrectly, to be dangerous to people; cure was said to be possible only if the victim danced the tarantella. Other spiders, called mygalomorphs but sometimes named tarantulas in South America, may have a span of 50 mm and can kill small birds.

Echinodermata

Echinoderms are radially symmetrical as adults: they have five radii (are pentaradiate); a typical starfish has five arms. But the larvae are bilaterally symmetrical and look very different. The adults are armoured and move by large numbers of tube feet—projections equipped with suckers. Starfish, sea urchins and others.

The Chordata have two subdivisions, listed below unconventionally as if they were separate phyla. They all have a dorsal nerve cord, a notochord and segmented musculature. So have we (at an early stage).

Cephalochordata (Acrania)

These simple chordates form a tiny group today, but they are represented among very early (Cambrian) fossils. *Branchiostoma* (or *Amphioxus*) is well known. They have a segmented musculature like that of a fish, a notochord and a nervous system formed from a dorsal neural tube, but no brain or distinct head. The gill clefts penetrate from the pharynx to the outside and are part of the equipment for filter feeding. They live in the sea and often burrow in sand.

Vertebrates (Craniata)

The vertebrates have a distinct head, a well developed brain in a cranium usually of bone, a vertebral column and four limbs. The head has eyes and other sense organs. The heart is ventral. The basic vertebrate is an active swimmer and respires by means of gills between the pharynx and the outside. The embryos of all land vertebrates retain traces of this arrangement.

JAWLESS VERTEBRATES (AGNATHA) have no jaws or paired fins. The lampreys and hagfish. Early forms, now long extinct, were often large and heavily armoured. Today, lampreys make luscious eating and notoriously encourage excess.

The remaining vertebrates have jaws.

CARTILAGINOUS FISH (CHONDRICHTHYES) have an internal skeleton of cartilage. Sharks, dogfish, skates and rays. The largest, the whale shark, *Rhineodon*, widespread in warm, surface waters, can be

18 m long and weigh 40 tonnes. Human swimmers below the surface have occasionally encountered it. They were ignored: the whale shark feeds by filtering small crustaceans and other plankton from the water. The great white shark, *Carcharodon*, often 6 m long and sometimes nearly twice that, can be a man-eater but is also a scavenger. The stomach of one, killed in Sydney Harbour, contained a tin can, many mutton bones, the hind quarters of a pig, the head and forequarters of a bulldog and some horseflesh, among other things.

BONY FISH (OSTEICHTHYES) have a skeleton of bone and a swim bladder (which, in a few, can function as a lung). Nearly all the familiar fish, and many unfamiliar ones. A lineage, separate from the rest, includes the few modern lungfish (Dipnoi) which have fins like those of the fish ancestral to the land vertebrates.

AMPHIBIANS lay eggs in water and usually have fish-like larvae (tadpoles) which breathe by gills; as adults they breathe with lungs but often also through the wet skin; they have two pairs of five-toed (pentadactyl) limbs. Newts, salamanders, frogs and many large extinct forms.

REPTILES are fully terrestrial and have a skin covered with hard scales; the eggs have a horny shell and are laid on land. Today we have crocodiles, lizards, snakes, turtles and a few others. Strange that the lumbering crocodiles have survived the magnificent, extinct creatures of the Mesozoic period (chapter 2). Also strange are the transformations from the original reptilian form: turtles and snakes hardly look as though they have evolved from ancestors resembling lizards. The turtles (Chelonia) are protected by a formidable dorsal carapace and, below, by a breast plate or plastron. The giants of this group, the size of a large bear, live on remote islands and, unfortunately, are highly edible. *Geochelone gigantea*, on Aldabra in the Indian Ocean, is a protected species. Perhaps some will therefore attain their reported span of more than 200 years.

Some snakes (Ophidia), too, make good eating despite their improbable anatomy. They are legless and manage with only one elongated lung. The forked tongue detects odours. The jaws of some can dislocate and allow the ingestion of prey of wider girth than the eater. Some snakes kill their prey by injecting poison through hollow fangs like hypodermic needles; others squeeze their food to death.

BIRDS (AVES), usually put in a separate group but regarded as dinosaurs by experts, are homeothermous ('warmblooded') and have feathers and hollow bones. The forelimbs are wings. They lay eggs covered with a hard shell. They see very well and some are now known also to have a good sense of smell. The constant body temperature fails only in the hummingbirds (Trochilidae) which weigh about 5 g and at night go into a cool torpor. During the day, hummingbirds suck nectar from flowers through long beaks. Their wings beat at up to 80/sec and enable them to hover and to fly backwards. The largest bird that flies, the wandering albatross (*Diomedea exulans*), has a wing span of 3 m, can glide for hours over the oceans without effort and is one of those able to sniff out the presence of food. Some, interfering with the US armed forces on Midway Island, were caught and released at a distant spot. They returned: one flew more than 5000 km, at an average of 507 km a day. The fastest, the swifts (*Apus*), include a species capable of 170 km/h. They spend nine months in each year airborne and touch down only when breeding.

Some birds are flightless. The penguins (Spheniscidae) have webbed feet but swim mainly with their forelimbs—wings which have become flippers. On land, flightless birds are or have been of several kinds. One, *Diatryma*, lived in Wyoming, USA, about 65 million years ago and reached over 2.0 m; its massive beak was probably used to prey on smaller animals. The flightless forms of today, evidently derived from flying ancestors, include the ostriches (*Struthio*) of southern Africa—the largest living birds. Among those of yesterday were the moas (*Dinornis*) of New Zealand, some more than 3.0 m tall but easily caught and eaten by human invaders. The dodo (*Raphus*) of Mauritius was a large pigeon which, like the moas, was killed off by hungry human visitors.

Some birds perform astonishing migrations. The legendary white stork (*Ciconia ciconia*) rears its young in central and southern Europe but spends the northern winter in Africa, south of the equator. Arctic terns (*Sterna macrura*) nest in the Canadian Arctic but avoid the Canadian winter by migrating to the Antarctic. The Pacific golden plover (*Pluvialis dominica*) breeds in Siberia and Alaska but spends the northern winter in South East Asia or Pacific islands. Migration by some species is guided by the patterns of the stars in the night sky and by the earth's magnetic field.

MAMMALS are homeothermous, have hair and are viviparous (monotremes excepted). The females feed their young with milk secreted by modified sweat glands. They have the largest brains in the animal kingdom.

Monotremes (platypus and echidna) lay eggs. Marsupials (kangaroos and many others) have yolky eggs and only a rudimentary placenta; the young are born very immature and develop in a pouch (marsupium). Placentals have a placenta and small eggs with no yolk.

The smallest mammals, the shrews (Soricidae), may weigh only a few grams; they have to spend most of their time eating. The Insectivora, the order to which shrews belong, have a complete array of teeth: cutting incisors; piercing canines; grinding premolars and molars. The largest living land mammal, the African elephant (*Loxodonta africana*), weighs in at 7.0 tonnes. Elephants have only tusks and massive grinders.

The brainiest mammals are intermediate in size. The Primates, tree shrews, lemurs, monkeys and apes include only about 170 species. They have binocular vision and are exceptionally dexterous. For more about Primates, especially *Homo sapiens*, see chapters 7, 9 and 11.

Appendix II

A Stratification of Organisms: The Record of the Rocks

The Main Geological Periods

Period		Million Years Ago
QUARTERNARY	Emergence of the human species.	1–2
TERTIARY	Flowering plants, insects, mammals and birds. *[See separate chart on next page.]*	75
CRETACEOUS	The last age of reptiles. Flowering plants, modern insects and mammals become more numerous; marsupials appear.	145
JURASSIC	Gigantic dinosaurs (as well as many small ones); large marine reptiles; pterodactyls; early mammals and birds; many bony fish; first flowering plants and hardwood forests.	190
TRIASSIC	Dinosaurs begin; forests of large conifers; first mammals.	240
PERMIAN	Reptiles and insects become diverse; first conifers.	290

Period		Million Years Ago
CARBONIFEROUS	Amphibians diversify; first reptiles, insects and seed plants; many sharks and bony fish; giant ferns and other spore-bearers; coal formed.	350
DEVONIAN	First amphibians; first land forests of ferns and others; armoured fish.	410
SILURIAN	Early fish, including ancestral lungfish; first land plants and animals.	430
ORDOVICIAN	First vertebrates (fishlike but jawless); many molluscs, algae.	500
CAMBRIAN	A large array of marine invertebrates appears 'suddenly'; plenty of algae.	600
PRECAMBRIAN	Microorganisms such as bacteria; small, obscure, soft-bodied animals.	>600

Each period is subdivided by geologists; subdivisions of the Tertiary follow. All estimates of duration are very approximate.

Mammals of the Tertiary and Quarternary Eras

Epoch		Duration (million years)
Recent	Neolithic society; agriculture—and all that has followed.	0.1
Pleistocene	Large mammals, including *Homo*, prominent. Cold climate.	1–2
Pliocene	Great diversity of mammals; early Hominoidea ('manlike' apes)	11
Miocene	Grazing mammals become numerous; early apes diversify.	15
Oligocene	Modern families of mammals, including early apes and monkeys.	11
Eocene	Modern orders of mammals, including early horses, whales, bats.	20
Paleocene	Ancient mammalian forms, including early Primates.	15

During all this time birds of the familiar groups were present; to say nothing of all the other plants, fungi, animals and microorganisms.

Appendix III

Glossary

ABIOGENESIS 'Spontaneous generation' or the development of living from nonliving things. No longer accepted as possible (except in the laboratory?). But living things are assumed to have arisen from inanimate matter in the remote past.

ADAPTATION In biology has two important meanings. (1) Physiological (ontogenetic) adaptation is a change in a single individual which enables it to cope better with its environment: for instance, acquiring resistance to disease; or enlargement of muscles with exercise; or learning the way about. (2) Genetical or phylogenetic adaptation occurs when a population changes genetically and so increases the chances of its survival.

ADIPOSE TISSUE A tissue in which most of the cells can store much fat.

AEROBIC RESPIRATION See respiration.

AGGRESSION A word confusingly applied to many kinds of violence, even to 'threats' and to defensive territorial warnings such as bird song. Not recommended for serious descriptions of animal or human behaviour.

ALLELE A gene (q.v.) is said to be an allele (or alternative) of another gene when it occupies the same position (locus) on a chromosome but produces a different effect on development. One allele can mutate into another. See mutation.

ALTERNATION OF GENERATIONS In a life cycle, a generation which reproduces sexually alternates with one which reproduces asexually. The two generations are usually very different. Coelenterates (jellyfish and others) are examples among animals. Among plants, mosses and ferns have a sporophyte (q.v.) which is diploid (q.v.) and reproduces asexually by spores (q.v.); and a gametophyte, which is haploid and reproduces sexually. Flowering plants have concealed vestiges of such alternation.

AMINO ACIDS The chemical units which, joined together, make proteins. They are small molecules: each contains basic amino (NH_3) and acidic carboxyl (COOH) groups. Twenty amino acids are important. Some are essential components of the human diet and must be present in the proteins in a person's food.

ANAEROBIC RESPIRATION Liberation of energy in a cell by breaking down substances without consuming oxygen. See respiration.

ANTIBODY A protein produced in response to the presence of an injurious substance (an antigen, q.v.) in the body. A defense mechanism especially important for resistance to infection by bacteria or viruses. See also immune response.

ANTIGEN A substance which provokes the formation of an antibody (q.v.).

APOSEMATISM Conspicuous appearance associated with an aversive taste or toxicity. A means by which some animals, especially insects, are protected against predators. See also mimicry.

ASEXUAL REPRODUCTION Reproduction without gametes (q.v.). Plants and protists may reproduce asexually by producing spores (q.v.) and protists by dividing; plants may reproduce vegetatively (as when a gardener propagates by cuttings); some animals reproduce by budding (for instance coelenterates such as *Hydra*).

AUXIN A plant hormone or growth factor.

AXON A long process of a nerve cell which normally conducts impulses away from the cell body.

BACTERIUM (plural: bacteria) Bacteria are microorganisms, nearly always single cells, without nuclei (prokaryotes, q.v.). Essential in soil formation and in maintaining the cycles of carbon, nitrogen and other elements. Many are parasites; some cause human diseases.

BEHAVIORISM The doctrine that the proper subject of scientific psychology is behaviour only.

BIOSPHERE The whole assemblage of living things on earth and the environments that support them.

BLASTULA (plural: blastulae) Stage in embryonic development. Typically, a hollow ball of cells; but when the egg contains much yolk this form is altered (as in the development of a bird).

CARBOHYDRATE A large class of organic compounds containing carbon, hydrogen and oxygen. Glucose, one of the many sugars, has the formula $C_6H_{12}O_6$; sucrose, the sugar on our tables, is $C_{12}H_{22}O_{11}$. Other carbohydrates, with much larger molecules, are starch and cellulose. Sugars and starch are important sources of energy in food; but cellulose is not digestible by human beings.

CARBON CYCLE The circulation of carbon atoms in the biosphere (q.v.). Carbon dioxide from air or water is made up into complex substances by plants (see photosynthesis); these substances are broken down in two main ways: (1) during the respiration (q.v.) which goes on in the plants themselves and in animals which eat the plants; and (2) by decay, mainly in soil or in seas and fresh waters. The breakdown releases carbon dioxide into the air and the waters. Compare ecosystem.

CELL Unit of living matter, usually microscopic, bounded by a very thin plasma membrane, and in plants also by a wall of cellulose. Usually contains a nucleus (q.v.; see also eukaryote); but the smallest organisms (bacteria [q.v.] and others) are cells without nuclei (prokaryotes [q.v.]). See also cytoplasm. The largest cells are very yolky ova (q.v.) such as those of birds.

CHLOROPHYLL Green pigment in cells of all the organisms in which photosynthesis (q.v.) takes place; makes possible use of light energy in synthesis of complex substances from carbon dioxide and water.

CHLOROPLAST An organelle (q.v.) containing chlorophyll (q.v.). The site of photosynthesis (q.v.) in plant cells.

CHROMATID One of the two strands resulting from duplication of a chromosome (q.v.); visible during the early stages of mitosis (q.v.); the strands later separate to become daughter chromosomes.

CHROMOSOME A body, usually thread-like, consisting mainly of DNA (q.v.) and protein (q.v.). Chromosomes make up most of the contents of the nucleus (q.v.) of a plant or animal cell (q.v.). In the somatic cells of animals, that is, cells which are not gametes (q.v.), and in the cells of plant sporophytes (q.v.), chromosomes occur in pairs (the diploid condition); the members of a pair are identical in appearance under an ordinary (light) microscope and are said to be homologous. Gametes have only one chromosome of each pair (the haploid condition). The number of chromosomes varies with the species. Chromosomes are easily seen only during nuclear division.

CLONE Individuals of which all have been produced by asexual reproduction (q.v.) from a single individual. In the absence of mutation (q.v.), all have the same genetical constitution.

CODON See gene.

COMMENSAL Members of different species living in close association, but not in symbiosis (q.v.) or parasitism; literally, 'sharing the same table'. Examples are the rats and mice associated with human beings.

CONDITIONAL REFLEX A response or act, elicited by a previously indifferent stimulus, as a result of the repeated occurrence of the indifferent stimulus at about the same time as an existing stimulus for a similar act. The classical example is from Pavlov's dogs: a buzzer is sounded before food is offered; at first, the buzzer attracts attention but has no effect on salivation; but, after many repetitions, salivation occurs in response to the buzzer before food appears.

CONGENITAL The primary meaning is: evident or detectable at birth. Sometimes misused to mean the same thing as genetical or genetic (q.v.).

CRYPTIC COLOURING Resembling the background. A means by which some animals escape the attentions of predators. Contrast aposematism.

CYANOBACTERIA 'Blue-green algae'. Microorganisms, capable of photosynthesis, which may resemble those which, in the remote past, were largely responsible for the greening of the biosphere (q.v.). Today, they are among the first to colonise bare rock (see succession). They also form toxic scums in inland waters.

CYTOPLASM All the protoplasm, or living material, of a eukaryotic cell, except the nucleus (q.v.). See eukaryote.

DENDRITE Short, branching projection of a nerve cell. Makes connections (synapses, q.v.) with the dendrites or axons (q.v.) of other cells.

DENSITY-RELATED FACTOR (ecology) Anything that tends to reduce the growth rate of a population of organisms and acts with increasing strength as the population density rises. The most important are food supply, predators, infectious organisms, shelter and crowding. See feedback.

DEOXYRIBONUCLEIC ACID See DNA.

DIPLOID Having the chromosomes (q.v.) in pairs: twice the haploid number. Most animal cells, except the gametes (q.v.), are diploid. See alternation of generations.

DNA Deoxyribonucleic acid. A giant molecule consisting of many nucleotides forming a chain; usually two chains are joined, parallel to each other, and coiled in a helix. Each nucleotide contains one of four bases (thymine, cytosine, adenine or guanine) and a sugar, deoxyribose. Found mainly in the chromosomes (q.v.) of animals and plants, and in corresponding structures of bacteria (q.v.). The order in which the bases occur in the chain is the 'genetic code' which, with the intervention of RNA (q.v.), determines the synthesis of amino acids (q.v.) and so of proteins. DNA is reproduced whenever a nucleus divides. The material basis of biological inheritance. See also gene.

DOMINANCE (ethology) See status system.

DOMINANCE HIERARCHY See status system.

DOMINANT (adj.) (genetics) Applied to a trait appearing in the phenotype (q.v.) as the result of the presence of a single copy of a gene (q.v.). In the corresponding position (locus) on the homologous chromosome (q.v.), a different form (an allele, q.v.) of the gene may be present; this is the heterozygous state and the individual is said to be, in respect of this locus, a heterozygote. When a trait appears only if two identical copies of a gene are present (the homozygous state), one on each of two homologous chromosomes, that trait is said to be recessive. In some writings, genes (as well as traits or characteristics) are called dominant or recessive. This usage is not recommended: the effects of genes are often multiple; some of the traits they influence may then be dominant but others, recessive.

ECOLOGY The science of the relationships of organisms with their environment, including the environment provided by other organisms. An ecologist may study associations of species, as in a rock pool, a patch of soil, a desert, a wood or the human skin; or the changing numbers and density of a population of organisms; or the flow of matter and energy through an association of species; and much else.

ECOSYSTEM An association of interacting organisms, together with their nonliving environment. Such a system includes producers (green plants) which synthesise organic matter from inorganic; consumers (mainly animals); and decomposers (mainly bacteria and fungi).

ENDOCRINE ORGAN An organ which secretes hormones (q.v.).

ENZYME A substance, usually a protein, which promotes (catalyses) a specific chemical change. The change may take place inside a cell or in a fluid secreted by cells (for instance, the digestive fluid in an animal's gut). Enzymes play an essential part in the metabolism (q.v.) of all organisms.

EPIGENESIS The appearance of new structures during individual development (ontogeny); this entails an interaction of the effects of genes with environmental influences. Opposed to the obsolete doctrine of preformation, according to which the organism is already fully formed in the fertilised egg or even in a gamete (q.v.).

EPIPHYTE A plant that grows on another plant which provides only support: examples are many lichens and mosses; also climbers such as ivy (*Hedera helix*).

ETHNOBOTANY The study of the relationship between plants and human beings, especially 'indigenous' people who are not technically advanced.

ETHOLOGY The science of animal behaviour. A major division of biological science, like ecology (q.v.). In some writings (but not this book) used in a narrower sense—the study of the behaviour of animals in their natural surroundings.

EUGENICS Maintaining or improving the quality of a human population by controlled breeding.

EUKARYOTE Of cells (q.v.) or organisms: having a nucleus (q.v.) separated from the rest of the cell (the cytoplasm) by a membrane. Contrast prokaryote.

EVOLUTION, ORGANIC The descent of organisms from very different organisms in the past. Traceable in the fossil (q.v.) remains of organisms. Evolutionary change is still going on; but it is very slow and can be observed in a human lifetime, if at all, only on the smallest scale.

EXPLORATION In ethology (q.v.), movements of an animal about its living space which are independent of any special need, for instance for food. They represent a tendency to approach strange objects and places and are a means by which an animal learns about its surroundings.

FAT A class of substances composed of glycerol and fatty acids; in animals, stored in fat cells (adipocytes). An important source of the energy provided by some foods. The word 'lipid' is sometimes used with the same meaning. Fat is not the same as adipose tissue (q.v.).

FEEDBACK Transfer of output to input, in such a way as to modify the input. In negative feedback the effect is to reduce the input; in positive, the input is enhanced. Negative feedback is universal in living systems, from the regulation of metabolism (q.v.) within cells to the control of the growth of populations.

FITNESS In biology, the fitness (also known as Darwinian fitness) of an organism is some measure of its contribution to later generations. There is no necessary connection with athletic prowess.

FLAGELLUM (plural: flagella) Fine, long, movable thread, projecting from a cell (q.v.). Most spermatozoa (q.v.) have one. The flagellate protozoans each have at least one.

FOSSIL Petrified remains of organisms in rocks. During petrifaction, spaces or pores in the tissues of dead organisms are filled with minerals. Usually only hard parts, such as bones and shells, are preserved in this way.

GAMETE Haploid (q.v.) cell of which the nucleus can fuse with that of another cell to produce a diploid zygote (q.v.). Most gametes are either mobile spermatozoa produced by a male organism, or much larger, immobile ova produced by a female.

GAMETOPHYTE The stage in the life cycle of a plant in which the cells have haploid nuclei. Produces gametes (q.v.).

GASTRULA (plural: gastrulae) A stage in animal development with two or three layers. Migration of cells of the blastula (q.v.) produce an outer ectoderm and an inner endoderm and, later, a third layer, the mesoderm, between the first two.

GATHERER HUNTERS People without agriculture or herds, who live by gathering and hunting food.

GENE Originally meant a unit of heredity of which the existence was inferred from breeding experiments. The units are passed on from parent to offspring unaltered. Soon after 1900 (when Mendel's findings were rediscovered), the genes or hereditary factors were shown to be arranged in line in the chromosomes (q.v.) of the cell nucleus (q.v.). Later, the material of heredity was found to be DNA (q.v.). Hence today the word gene usually means a length of DNA. The nucleotides (q.v.) in the DNA molecule are arranged in a linear sequence. Each group of three nucleotides (codon) codes for an amino acid. A gene consists of many similar codons. Much chromosomal DNA is, however, 'noncoding': see transposable element.

GENETICAL (also GENETIC) Usually means related to the action of genes. May properly refer to *differences* between organisms: some differences are genetically determined. Characteristics or traits are sometimes said to be genetical but this usage is not recommended: *all* traits are influenced by both the genes *and* the environment.

GENETIC CODE See DNA.

GENETICS The science of variation and heredity.

GENOME The whole array of the genetical material (DNA, q.v.) of a species.

GENOTYPE The genetical constitution, that is, the whole set of genes, of an organism. Contrast the phenotype, which is its whole set of actual characteristics.

GEOTROPISM Regulation of the growth of parts of plants by gravity. In the dark, main stems of flowering plants grow upward (negative geotropism); and roots, downward (positive geotropism). See also auxins.

GERM CELL The same as gamete (q.v.).

GOLGI BODY An organelle (q.v.) in the cytoplasm (q.v.) of animal and plant cells. In electron micrographs appears as a group of flat sacs plus separate vesicles. Especially prominent in cells which produce secretions.

GONAD An organ of an animal in which gametes (ova or spermatozoa) develop. The female gonad is an ovary; the male, a testis. Gonads of many animals also secrete hormones.

GREENHOUSE EFFECT Much of the heat of the sun is reflected off the earth's surface; but some is retained, as in a greenhouse, because it is reflected back again by a layer of gases, such as carbon dioxide and methane, in the upper atmosphere. The expression, greenhouse effect, today usually refers to a *rise* in the earth's temperature due to the current increase in the amount of 'greenhouse gases'.

HAPLOID Having a single set of unpaired chromosomes. Gametes (q.v.) are haploid; so are the cells of gametophytes (q.v.). An organism consisting of such cells may itself be said to be haploid.

HERITABILITY A measure of the extent to which, in a specified population, genetical differences contribute to the observed phenotypic variation. A high heritability does *not* signify that the trait measured is little affected by environmental change. See also phenotype.

HETERODONT Possessing teeth of several kinds. Characteristic of mammals. Contrast those of reptiles, all of which are pegs or spikes.

HISTOLOGY The scientific study of the tissues of plants, fungi and animals.

HOMEORHESIS Maintenance of a constant sequence of developmental stages.
HOMEOSTASIS Maintenance of a steady internal state: for instance, blood composition.
HOMEOTHERMY (homothermy, homoiothermy; also endothermy). Maintenance of a constant body temperature within a range of external temperatures. Characteristic of birds and mammals. An aspect of homeostasis (q.v.).
HOMOLOGY Similarity in structural relationships, especially in embryonic development, regardless of function. The main components of the human forelimb, a whale's flipper and the wings of a bat or a bird are all homologous. The similarities are attributed to evolution from a remote common ancestor.
HORMONE Organic substance secreted in very small amounts in one part of an organism and carried to another part where it influences metabolism, growth or other processes.
HUNTER GATHERERS Alternative name for gatherer hunters (q.v.).
IMMUNE RESPONSE A response to infection which protects an organism and leaves it less susceptible to the infective agent. See also antibody, lymphocyte, lymphoid tissue, phagocyte.
IMPRINTING Development, during an early sensitive period, of the tendency to follow or approach a particular object, usually a parent. When birds or mammals are active soon after hatching or birth, they perform this 'following response'. In doing so, they learn the special features of the parent.
INSTINCT (1) In everyday speech, often means the same as intuition or unconscious skill: as when a person is said to avoid a blow instinctively or to understand the attitude of another person by instinct. (2) When an animal performs a complex act without learning how to do it, this may be called instinctive behaviour. In ethology (q.v.), this usage is being given up, for reasons outlined in chapter 9. (3) A third meaning is the same as, or similar to, that of drive or the impulsion to act in certain ways. In some writings expressions like hunger drive, aggressive drive and so on still occur. This usage, too, is being given up: instead, the actual behaviour is described and, when known, the physiology underlying the behaviour.
KIN SELECTION Natural selection (q.v.) is held to favour traits that help both an individual's offspring and also its near relatives.
LAMARCKISM In ordinary speech, the statement that acquired characters are inherited. It is still sometimes assumed that the effects of use and disuse of organs (such as muscles), and of practice, are transmitted from one generation to the next. On this view, apparent skills, such as those of a bee building a comb or a predator tracking and killing prey, are inherited memories, derived from ancestors which had to learn these skills. We have no evidence that such transmission can happen. Our knowledge of the way in which DNA (q.v.) operates contradicts it.

LYMPHOCYTE A type of small white blood cell of vertebrates. Produces antibodies.

LYMPHOID TISSUE A vertebrate tissue which produces lymphocytes (q.v.); occurs especially in thymus, lymph nodes and spleen.

LYSOSOMES Very small granules in the cytoplasm of animal cells; contain enzymes, liberated when cells are damaged, which help to digest and remove the remains of dead cells.

MEIOSIS Also called reduction division. Two successive nuclear divisions during which the chromosomes (q.v.) divide only once. The initial nucleus is diploid. The four nuclei that result are haploid. Occurs at some stage in the life cycle of all sexually reproducing organisms, usually in the formation of gametes (q.v.). At fertilisation, the diploid number of chromosomes is restored. See alternation of generations. ('Meiosis' is sometimes used for the type of *cell* division in which these changes take place.)

MERISTEM A group of undifferentiated plant cells capable of division. These cells differentiate into specialised cells. In flowering plants meristems occur at the tips of growing stems and roots, in young leaves, in bark and elsewhere.

METABOLISM The chemical changes which go on all the time in an active, living organism. They include the breaking down of substances with release of energy (respiration, q.v.) and the building up of complex substances from simple ones.

METAMERISM Segmentation. In many animals, especially annelids (earthworms and others) and arthropods (insects, crustaceans and others), the main kinds of organs occur repeatedly, from front to rear, in separate sections or segments of the body.

MIMICRY Similarity of one species of animal to another, with a protective function. Sometimes, one species is both poisonous or distasteful and conspicuously marked, while the other (the mimic) is only conspicuous: this is called Batesian mimicry. Sometimes, however, both species are toxic and conspicuous: this is Mullerian mimicry.

MITOCHONDRION (plural: mitochondria) A type of organelle (q.v.) occurring in the cytoplasm of all cells except bacteria (q.v.) and cyanobacteria (q.v.). Contains many enzymes (q.v.) which catalyse the breakdown of substances with release of energy.

MITOSIS Division of a nucleus into two, with production of two daughter nuclei, each with the same number of chromosomes. Not the same as cell division.

MUTANT See mutation.

MUTATION Sudden change in the DNA (q.v.) usually of a chromosome (q.v.). Mutations can occur in any cell nuclei. Those that occur in a gamete produce mutant DNA which can be passed on to later generations. Of these, the most important are changes in single genes (or codons, q.v.). DNA is very stable: mutation is a rare event. It can, however, be speeded up by radiation and by some poisons. Most mutant genes have an adverse effect on the organism; but some, especially in a changing environment, may be advantageous. See natural selection; nucleus; transposable element.

MYCORRHYZA Fungus root. Symbiotic association of a fungus with a plant. See symbiosis.

NATURAL A word of many meanings. The 'natural environment' of a species means, as a rule, the environment in which the species usually lives and so the one to which it is adapted. Similarly, the expression 'in nature' commonly means in normal or usual conditions. 'Natural' is then opposed to 'artificial'. Problems arise when this distinction is applied to human affairs. Early human beings lived as gatherer hunters (q.v.); this condition may then be said to be natural for the human species. Yet it is also 'natural' (normal, usual) for human beings to change the conditions in which they live (especially by using tools which themselves change with time); people also domesticate and alter the plants and animals on which they depend. Hence nearly all our food and other organic materials are 'artificial': they are products of human ingenuity. In advertisements and other propaganda, 'natural'—if it has a meaning—signifies 'good'. The question then is, good for what?

NATURAL SELECTION The name for a process in nature which results from the existence of genetically determined variation among organisms. As a result of this variation, some organisms contribute more to later generations than do others. They are then said to have greater fitness (q.v.) than the others. Evolutionary change is held to be largely due to differences of fitness in this sense. But natural selection can also *prevent* change: see stabilising selection.

NATURE Like 'natural' (q.v.), a word of many meanings; not a technical term in biology. Here is a comment by an art historian, Kenneth Clark:

> We are surrounded with things which we have not made and which have a life and structure different from our own: trees, flowers, grasses, rivers . . . They have inspired us with curiosity and awe. They have been objects of delight . . . And we have come to think of them as contributing to an idea which we have called nature.

NEURON A nerve cell. Has processes projecting from it, often very many. The cells and their processes conduct impulses among themselves; or from sense organs to a central nervous system; or from the central nervous system to muscles or glands. Transfer of impulses from one neuron to another is at junctions called synapses (q.v.). See also axon.

NITROGEN CYCLE The circulation of nitrogen atoms in the biosphere (q.v.). Soluble, inorganic nitrogen compounds, mainly nitrates, are taken by plants from soil or water and synthesised into complex organic substances, especially proteins (q.v.). These either return to the soil or waters; or they are eaten by animals which, in turn, pass them back to soil or water in excreta or by death. There, bacteria convert them again to simple inorganic sustances. Some nitrogen is lost to the air; but some atmospheric nitrogen is fixed by nitrogen-fixing bacteria. The maintenance of this and other cycles of matter in the biosphere is crucial for the survival of nearly all life. In human affairs it is aided by measures such as rotation of crops and manuring fields. Compare carbon cycle, ecosystem.

NUCLEIC ACID See nucleotide.

NUCLEOLUS Small body, within a nucleus (q.v.), consisting mainly of DNA (q.v.) and protein (q.v.), which disappears during mitosis. It is concerned with the production of the RNA (q.v.) which intervenes between DNA and the formation of amino acids.

NUCLEOTIDE A class of substances formed from a sugar, phosphoric acid and a nitrogen-containing base. Nucleic acids are made up of nucleotides. See also DNA, RNA, both of which are nucleic acids.

NUCLEUS The part of a cell which contains the chromosomes (q.v.). Present in nearly all the cells of many-celled organisms. See eukaryote.

ORGANELLE A distinct part of a cell (q.v.), with a special function. Analogous to the organs of many-celled organisms.

OVUM (plural: ova) Female gamete (q.v.). Ova are larger than most other cells; some are very large. See also zygote.

OZONE LAYER Ozone is a form of oxygen (O_3) which is present in the upper part of the earth's atmosphere (the stratosphere). It absorbs much of the ultraviolet (UV) radiation from the sun. In doing so, it protects organisms (including human beings) from damaging radiation. It also yields some heat which contributes to the stability of the atmosphere. The stratospheric ozone is being depleted as a result of pollutants released into the air. The most important pollutants are the chlorofluorocarbons, the use of which is slowly being reduced. In the 1990s, ozone depletion is nevertheless increasing at an alarming rate, in both hemispheres.

PECK ORDER A term for a kind of status system (q.v.).

PHAGOCYTE A cell which, like an ameba, takes in particles from its surroundings. Most many-celled animals have phagocytes in their body fluids, where they protect the body from invading microorganisms.

PHENOTYPE All the characteristics of an organism. Contrast genotype (q.v.). Two organisms can have the same genotype, but different phenotypes, because their environments are different. See epigenesis.

PHEROMONE Odorous substance that acts as a social signal.

PHOTOSYNTHESIS Synthesis of organic substances from carbon dioxide and water by use of energy absorbed by chlorophyll from sunlight. Takes place inside chloroplasts (q.v.). The existence of all organisms on earth, except a few kinds of bacteria, depends directly or indirectly on this process.

PHOTOTROPISM Growth of part of a plant toward light. Depends on the growth 'hormones' called auxins (*q.v.*).

PLACENTA The afterbirth. An organ, consisting of both embryonic and maternal tissues, by which an embryo is supplied with food and relieved of waste products. Animals which reproduce in this way are called viviparous.

POIKILOTHERMY (ectothermy) Having a body temperature which varies with that of the surroundings. Contrast homeothermy.

POLYANDRY Mating by a female with more than one male during a single breeding season. Also applied to possession by a woman of more than one husband. Compare polygyny.

POLYBRACHYGAMY Mating with two or more others during the same breeding season. The human equivalent is promiscuity.

POLYGYNY Mating by a male with more than one female during one breeding season. A human parallel is possession of a harem.

POLYMER A substance built up from a series of smaller chemical units (monomers). Proteins (q.v.) and nucleic acids (q.v.) are polymers, built from amino acids (q.v.) and nucleotides (q.v.), respectively.

PREFORMATION See epigenesis.

PROKARYOTE Possessing genetical material as filaments of DNA which are not separated in a nucleus. Bacteria (q.v.) and cyanobacteria (q.v.) are prokaryotes.

PROTEIN Complex organic substance (a polymer, q.v.) made up of large numbers of amino acid (q.v.) units. Present in all living things and only in them. Each species of organism has its own characteristic proteins.

RECAPITULATION In a narrow sense, means the appearance in embryonic development of features resembling those of ancestral adults. It was once thought that the embryonic stages of, for example, mammals (including human beings) correspond accurately to their evolutionary history. They do not. But the developmental stages of all land vertebrates (which have evolved from fish) do resemble the early stages of a fish: the heart first has a structure like that of a fish; the muscles are in similar blocks or segments; there is a muscular tail; gills begin to develop; and so on.

RECESSIVE (genetics) See dominant (genetics).

REDUCTION DIVISION See meiosis.

REDUCTION, EXPLANATORY The findings of one kind of study, such as genetics, can sometimes be partly explained by reduction to those of another, such as biochemistry. So knowledge of the chemistry of DNA (q.v.) helps us to understand the phenomena of heredity. Another kind of example is the attempt to explain human social action in terms of genes and the presumed past action of natural selection (q.v.). The power of reduction, especially in the physical sciences, has led some people to believe that all biology can be explained by physics and chemistry or even in terms of mechanisms. But, if organisms and human beings are to be understood, it is first necessary to make nonreductionist statements about organisms and human beings. These cannot be replaced by biochemical, genetical or other reductionist statements.

REFLEX An immediate reaction to stimulation, highly predictable, performed always by the same muscles or glands and consistently related to a particular kind of stimulation. Examples include a sneeze in response to something in the nose; the secretion of saliva in response to something in the mouth; and the contraction of the pupil of the eye in response to light.

REGULATION (in embryonic development) The development of a normal structure in spite of disturbance. For example, if, in the earliest stages, the cells of a developing sea urchin are separated, each can develop into a complete (though small) larva.

RESPIRATION Has several meanings. (1) Movements by which air is passed in and out of lungs (breathing) or water is passed over gills. (2) Taking oxygen from air or water and giving off carbon dioxide; these are processes which go on in lungs or gills. (3) Cellular respiration, the most fundamental, is the set of processes by which energy is released in cells. Usually it depends on the oxidation of a substance such as glucose, with release of carbon dioxide. This is aerobic respiration. The energy produced equals what would be released if the glucose had been burnt in air; but release is relatively slow: hence the cell does not burst into flames. In anaerobic ('airless') respiration, energy is liberated by breakdown of substances without the help of molecular oxygen. Some microorganisms, such as certain yeasts, can respire in either way. Some bacteria are exclusively anaerobic.

RIBONUCLEIC ACID RNA (q.v.).

RIBOSOMES Granules consisting of RNA (q.v.) and protein (q.v.) present in the cytoplasm of all cells; the site of protein synthesis.

RNA A giant molecule consisting of many nucleotides joined in a chain. Each nucleotide contains one of four bases (uracil, cytosine, adenine or guanine) and a sugar, ribose. Found largely in the ribosomes (q.v.) of animals and plants, and as messenger RNA and transfer RNA, both of which are concerned in translating the 'coded' information in DNA (q.v.) into protein structure.

SEGMENTATION See metamerism.

SEXUAL REPRODUCTION In biology, formation of a new individual as a result of the fusion of two haploid (q.v.) nuclei. Usually involves the fusion of two gametes (q.v.) to form a fertilised egg or zygote (q.v.).

SOCIAL Concerning interactions between members of the same species.

SOCIOBIOLOGY The interpretation of animal and human social behaviour by the supposed past action of natural selection. Applied to humanity, it exemplifies the limitations of explaining by reduction (q.v.).

SPECIES The smallest unit of classification (or taxon, q.v.) commonly used. All the members of a population assigned to one species, if they reproduce sexually, are usually assumed to be capable of interbreeding. When two kinds of organisms are assigned to different species, it is assumed that they *cannot* interbreed or that, if they do, they will not produce fertile offspring.

SPERMATOZOON (plural: spermatozoa) Small, active gamete (q.v.), usually with a flagellum (q.v.).

SPONTANEOUS GENERATION See abiogenesis.

SPORE Reproductive structure, usually microscopic, which becomes detached from a plant or a fungus, or is formed by a protozoan or bacterium, and can give rise to a new individual. A means of asexual reproduction (q.v.). Often produced in vast numbers and capable of surviving adverse conditions while in an inactive state.

SPOROPHYTE Stage in the alternation of generations (q.v.) of plants: has diploid (q.v.) cell nuclei (q.v.); produces spores (q.v.). Arises from zygote (q.v.) produced by union of sex cells which have developed in the gametophyte (q.v.).

STABILISING SELECTION If an organism is very different from the typical or normal for its species, it is likely to be less fit, in the biological sense (see fitness), than those nearer the norm. That is, in a constant environment, atypical forms tend to breed less, if at all, than the typical: they are 'weeded out'. When the differences are genetically determined, the typical form is preserved and the population is, in that sense, stable. But see natural selection.

STATUS SYSTEM In animal behaviour, a social system which includes relationships of dominance and subordinacy. The expression 'dominance hierarchy' is often used with the same meaning. The exact meanings of dominance and subordinacy depend on the observer and the species studied. Typically, a dominant individual has priority at place, food or a mate. Subordinates give way.

SUBORDINACY (ethology) See status system.

SUCCESSION (OF PLANTS) Sequence of changes in the composition of an association of plants. If a succession begins on bare rock, the first colonisers are likely to be cyanobacteria and lichens; these set going the slow process of soil formation. After many decades or centuries, and many intermediate stages, the outcome of a succession (sometimes called a climax) may be a forest.

SYMBIOSIS Association of members of two or more species to their common advantage.

SYNAPSE The functional connection of two nerve cells (or neurons, q.v.). 'Connection' does not signify continuity: a gap always remains between the surface membranes of the two cells. One cell influences another by secreting substances which pass across the gap. Most nerve cells are in synaptic relationship with many others.

TAXON (plural: taxa) A unit of biological classification. The principal taxa, from largest to smallest, are: kingdom, phylum, class, order, family, genus, species.

TEACHING In the present book, the word teaching means an activity that alters the behaviour of a member of the same species (the pupil) and tends to be persisted in until the pupil reaches a certain standard of performance or improvement. (In ordinary usage, 'teaching' has many other meanings.) The systematic teaching of skills is a distinctively human trait.

TERATOMA An abnormal growth containing a disorderly collection of tissues.

TERRITORY In animal behaviour, a region, occupied by an individual or a group of animals, from which other members of the species are excluded. Often reserved for a region *defended* from others. Different species have different sizes and kinds of territory. Territorial behaviour should be distinguished from behaviour which maintains a status system (q.v.) *within* a group of animals. It should also not be muddled up with human ownership of property.

TRANSFECTION The transfer of DNA (q.v.) between species. Allows the breeding or cultivation of forms with quite new features: for instance, a crop plant may be given resistance to disease derived from another, distant species.

TRANSPOSABLE ELEMENT (jumping gene; transposon) A mobile, virus-like particle, part of a DNA (q.v.) molecule which influences the development of an organism; but, unlike more familiar genes, it does not code for protein synthesis. It can alter its position on a chromosome or even move outside the cell. Seemingly can move from one individual to another, even to a different species. Transposable elements make at least 10 per cent of the genome (q.v.) of complex organisms. They are coming to be regarded as an important source of variation, in addition to mutation (q.v.), and as significant for our understanding of evolution (q.v.).

TROPISM Directed response of a plant to a stimulus, such as light or gravity. See geotropism, phototropism, for examples.

VIRUS Submicroscopic particle, consisting mainly of DNA (q.v.) or RNA (q.v.), which infects an organism and often causes disease. Viruses can multiply only in a living cell.

VIVIPARY The development of an embryo inside a female, until it has some capacity to survive outside. All placental mammals are viviparous. Vivipary is rare in the rest of the animal kingdom. See placenta.

X CHROMOSOME A chromosome (q.v.) which is paired in one sex but not in the other. Usually, the female is the sex in which the cells have two X chromosomes. The male's cells then have each only one X chromosome, which is paired with another, usually smaller, the Y chromosome.

Y CHROMOSOME See X chromosome.

ZYGOTE The fertilised ovum (q.v.).

FURTHER READING

The works listed have been chosen as sources of reliable information. Nearly all have good bibliographies.

Alberts B. et al. 1997 *Essential Cell Biology*, Garland, New York. An excellent and beautifully illustrated, elementary account of modern knowledge of cells. A model textbook.

Alexander R.M. 1990 *Animals*, Cambridge University Press, Cambridge. A concise textbook of zoology.

Balick M.J. & Cox P.A. 1996 *Plants, People, and Culture: the Science of Ethnobotany*, Scientific American, New York. The lavish illustration gives a coffee table appearance; but the lucid text contains important information on diet, drugs and conservation.

Barnes R.S.K. et al. 1988 *The Invertebrates*, Blackwell Scientific, Oxford. A readable textbook.

Barnett S.A. 1988 *Biology and Freedom: an Essay on the Implications of Human Ethology*, Cambridge University Press, Cambridge. Human biology and politics: misleading biological portraits of humanity and an alternative.

Basalla G. 1988 *The Evolution of Technology*, Cambridge University Press, Cambridge. The growth of technology in historical time contrasted with the evolution of organisms in geological time.

Campbell B.G. 1985 *Human Evolution*, Heinemann, London. A balanced survey for the serious reader.

Clutton-Brock J. (ed.) 1989 *The Walking Larder*, Unwin Hyman, London. Articles on human relationships with domestic animals.

Cott H.B. 1940 *Adaptive Coloration in Animals*, Methuen, London. A superbly illustrated book of great scope.

Crick F. 1988 *What Mad Pursuit*, Basic Books, New York. An enjoyable autobiography by the principal discoverer of the 'genetic code'. No references to sources.

Crosby A.W. 1986 *Ecological Imperialism: the Biological Expansion of Europe, 900–1900*, Cambridge University Press, Cambridge. The effects, often calamitous, of transferring species, including microorganisms and large mammals, from one region to another.

Desowitz R.S. 1991 *The Malaria Capers*, Norton, New York. An account, partly autobiographical, of research on tropical diseases, especially malaria, and of the follies and crimes which have obstructed it. The frivolous title is misleading.

Djerassi C. 1992 *The Pill, Pygmy Chimps, and Degas' Horse*, Basic Books, New York. An attractive autobiography by one of the principal creators of the contraceptive pill. The author has lived an exceptionally full life. No references to sources.

Dressler D. & Potter H. 1991 *Discovering Enzymes*, Freeman, New York. An introduction to biochemistry through the study of enzymes; very good on history.

Dréze J. & Sen A. 1989 *Hunger and Public Action*, Clarendon Press, Oxford. The vast scale of the problems, and some solutions.

Dubos R.J. 1951 *Louis Pasteur*, Gollancz, London. A wide ranging, absorbing biography of a genius, by a leading bacteriologist.

Eiseley L.C. 1959 *Darwin's Century*, Gollancz, London. Darwin and Darwinism in the nineteenth century. Very well written.

Elton C. 1958 *The Ecology of Invasions by Plants and Animals*, Methuen London. An agreeably conversational survey of ecological principles, illustrated by the effects of introduced species.

Feduccia A. 1980 *The Age of Birds*, Harvard University Press, Cambridge MA. A large book for the enthusiastic reader, on birds and their evolution.

Frisch K. von 1974 *Animal Architecture*, Harcourt-Brace Jovanovich, New York. A beautiful book for the general reader; no references to sources.

Futuyma D.J. 1986 *Evolutionary Biology*, Sinauer, Sunderland. An exceptionally good textbook.

Gould S.J. 1981 *Mismeasure of Man*, Norton, New York. A history of attempts to apply genetical and evolutionary ideas to the human species. Debunks many errors.

Gould S.J. 1991 *Wonderful Life: the Burgess Shale and the Nature of History*, Penguin, London. A lively account of the discovery and interpretation of early animal fossils; and of the discoverers and interpreters.

Haldane J.B.S. 1932 *The Causes of Evolution*, Longmans, London. Lectures addressed to a wide audience; still readable with pleasure and instruction, though out of date.

Harrison G.A. et al. 1988 *Human Biology*, Oxford University Press, Oxford. A textbook.

Hecht S. & Cockburn A. 1989 *The Fate of the Forest*, Verso, London. The social and political background of the destruction of the Amazon forests and of the struggles to prevent it. Horrifying but not without hope.

Keller E.F. 1983 *A Feeling for the Organism: the Life and Work of Barbara McClintock*, Freeman, New York. The absorbing story of an unpublicised geneticist who made a novel and outstanding contribution to biology.

Kevles D.J. & Hood L. (eds) 1992 *The Code of Codes*, Harvard University Press, Cambridge MA. Thirteen essays, technical, ethical or critical, on the Human Genome Project and related topics.

King A. & Schneider B. 1991 *The First Global Revolution*, Simon & Schuster, London. For once, the blurb is appropriate: the book 'is both a dreadful warning and an approach to a sustainable solution' to the problems arising from 'the headlong post-war period of economic growth in the industrialised world'.

King J. 1997 *Reaching for the Sun: How Plants Work*, Cambridge University Press, Cambridge. A cheerful and highly informative story, blessedly without gimmicks, mainly about flowering plant physiology—a globally important subject. But no pictures.

Kozol J. 1985 *Illiterate America*, Doubleday, New York. An exposure and a program of remedy.

Krebs C.J. 1988 *The Message of Ecology*, Harper & Row, New York. A short, reliable textbook.

Lewontin R.C. 1993 *Biology as Ideology: the Doctrine of DNA*, HarperCollins, New York. Broadcast lectures: constructive common sense on attempts to explain humanity by genes and natural selection.

Margulis L. & Schwartz K.V. 1988 *Five Kingdoms: an Illustrated Guide to the Phyla of Life on Earth*, Freeman, New York. An excellent summary for the serious student, with good pictures.

McMichael A.J. 1993 *Planetary Overload: Global Environmental Change and the Heatlh of the Human Species*, Cambridge University Press, Cambridge. A survey of public health in a biological context.

Montagu A. (ed.) 1984 *Science and Creationism*, Oxford University Press, Oxford. Articles on the debates between creationists and their critics.

Nelkin D. & Lindee M.S. 1995 *The DNA Mystique: the Gene as a Cultural Icon*, Freeman, New York. A highly readable exposure of the obsession with genes and genetics in human affairs.

Owen J. 1983 *Garden Life*, Chatto & Windus, London. An English garden, from January through December.

Passmore J. 1974 *Man's Responsibility for Nature*, Duckworth, London. A philosopher's historical survey of attitudes to the natural world: much valuable critical analysis, hesitant on solutions.

Ponting C. 1991 *A Green History of the World*, Sinclair-Stevenson, London. A foundation for understanding current problems.

Raven P.H. 1984 *Biology of Plants*, Worth, New York. A textbook; includes much on economic applications of botany.

Rose S., Kamin L.J. & Lewontin R.C. 1984 *Not in Our Genes: Biology,*

Ideology and Human Nature, Penguin, Harmondsworth. A popular example of reliable biology and common sense.

Rowell A. 1997 *Green Backlash: Global Subversion of the Environmental Movement*, Routledge, London. An account of ruthless, sometimes criminal attempts by the private sector to destroy or to discredit the movements to preserve the biosphere.

Schumann G.F. 1991 *Plant Diseases: their Biology and Social Impact*, APS Press, St Paul MS. An important subject, not always given enough attention; and, as this book shows, of absorbing interest.

Strachan T. & Read A.P. 1996 *Human Molecular Genetics*, Bios, Oxford. A lucid, critical account of human genetics. For any serious student who wants an authentic, up-to-date review of gene action and the interaction of heredity and environment.

Tansey G. & Worsley T. 1995 *The Food System*, Earthscan, London. A comprehensive summary of nutrition and the ecology, economics and politics of food.

Tuana N. 1993 *The Less Noble Sex: Scientific, Religious, and Philosophical Conceptions of Woman's Nature*, Indiana University Press, Bloomington. Largely on the fantasies, errors and confusions of European males writing about women during the past two and a half millennia. Revealing.

Tudge C. 1991 *Last Animals at the Zoo*, Hutchinson Radius, London. The extinction of animal species and how to prevent it by breeding in captivity; the formidable problems are undisguised.

Tudge C. 1993 *The Engineer in the Garden: Genes and Genetics*, Jonathan Cape, London. A user friendly introduction to modern genetics and its applications. References are scattered in the text.

Wallman J. 1992 *Aping Language*, Cambridge University Press, Cambridge. A detailed account of attempts to teach apes language, with an unusually clear exposition of the gulf between apes and ourselves.

Wilson E.O. 1992 *The Diversity of Life*, Harvard University Press, Cambridge MA. A well illustrated compilation of findings and proposals on biodiversity.

Young A.M. 1994 *The Chocolate Tree*, Smithsonian, Washington. An autobiographical account of research on the tropical cacao tree. This one species, of great economic importance, still presents biological problems.

The monthly journal, *Scientific American*, publishes excellent articles on many of the topics discussed in this book.

INDEX

abiogenesis 5–7, 93
adaptation 26, 204–6, 211–14
adipose tissue 103–4
adrenalin 100, 102
Ageleius phoeniceus 184
aggression 229
AIDS 43–6
albinism 64
Alexander, R. McNeill 16
alkaloids 35
alternation of generations 255
Alzheimer's disease 72, 73
Amoeba 87, 90
Annelida 260
Anomalocaris 15
antibiotics 46, 88, 256
    and natural selection 202
antibodies 39, 42
antivenin 39
Apatosaurus 16
Apis mellifera 176–9
Aptenodytes forsteri 180
aquaculture 158
Arber, W. 52
Archaeopteryx 25
Armadillo 111
Atropa belladonna 36
atropine 36
Australopithecus 224

baboons 182
bacille Calmette-Guérin (BCG) 46
bacteria 4, 87–9, 253
    and abiogenesis 7
    nitrogen-fixing 88
    symbiotic 89
Beagle 198–9, 210
behaviorism 132–6
Bernard, C. 99, 109
Binet, A. 76
biodiversity 167–9, 244
biological control 147, 161–3
birds 263–4
    see also named species
bird song 181, 184
bishop (bird) 180
bison 155
Biston betularia 200–1
Blakemore, C. 124
blood 33–4
blue-greens 88–9
bonobo 183, 219–21, 232
boredom 137
boubou shrike 181
bower birds 179–90
Brachiosaurus 16

brachydactyly 68
Bradshaw, A.D. 210
brain 124–142, 221, 224

and memory 130
breast feeding 115
Brenner, S. 53
Buceros rhinoceros 180
Buchner, E. 37
Bunyan, J. 31

cacao tree 36
caffeine 36
calicivirus 163
Calvin, W.H. 214, 232
cancer 5, 119–22
 of breast 72
 of lung 122
 and smoking 121
 uterine 120
Cannon, W.B. 100
carbon cycle 149–50
Carson, R. 242
Catholicism 95
'cave men' 227–8
cells 10, 31–57 *passim*
 differentiation of 117–19
 of plants 34–6
Cercopithecus aethiops 231
cerebellum 126
cerebral hemispheres 126–8
Chadwick, E. 241
chimpanzee 183, 190, 219–20
chlorophyll 35, 37
chocolate 36
Chondrodendron 36
chondrodystrophy 68
Christie, A. 138
chromosomes 10, 50, 63
 and sex 48
 and sex linkage 62
Cinchona 36
Clarke, A.C. 158–9
Clarke, E. 83
classifying organisms 19–21
Clever Hans 175
clone 115–17
 of cells 42, 117
 of DNA 52
colostrum 114
computers 138–40
 and garbage 71
conditional reflexes 131–4
conifers 151–2
Conium maculatum 36
cooking stoves 248

corals 258
Crick, F.H.C. 141
'Cro-Magnon man' 226
crowding 185–6
cuckoo 179, 187

Cuculus canorus 179, 187
curare 36
curiosity 136–7
 of chimpanzees 190
cyanobacteria 88–9
Cycliophora 260
cystic fibrosis 55, 69

Dacrydium franklini 239
Darwin, C. 25, 26, 198–9
 and human evolution 188
 and Lamarckism 197
 quoted 195
Darwin's finches 210
Dawkins, R. 141, 207
DDT 202
 danger from 241–2
 and malaria 92
deadly nightshade 36
deficiency diseases 107
definition 234
demographic transition 172
density-related factors 159–62
depression 74
Desowitz, R. 93
diabetes 78, 100
diatoms 91
Digitalis purpurea 35
dinosaurs 15–17
DNA 8–10, 50–4
 human 59, 219
dolphins 153, 159
domestication 64, 170–1
dominance (behavioral) 175, 182–3
dominance (genetical) 61, 68
dominance hierarchy *see* status system
double helix 50–1
Drepanididae 209
Drosophila melanogaster 61

Ebola virus 45
ecology 147–73
 human 237–52
ecosystem 149
ectrodactyly 68
eland 155
electron microscope 33
embryonic development 110–14
 and evolution 21–3
 human 114
emperor penguin 180

energy 245–6
enzymes 37–8
 engineered 89
 restriction 52
epigenesis 65–6, 73–85, 122–3
 and diet 104

epinephrine *see* adrenalin
eucalypts 152
eugenics 65–71, 199–200
Euglena 91
eukaryotes 9
Euplectes franciscans 180
eusociality of insects 174, 177–9
evolution 12–26
   causes of 195–215
exploration 132, 136–8
extinction (behavioural) 133
extinction (of species) 167–9
   of dinosaurs 18

farming 170–1, 237, 247
feedback, negative *see* negative feedback
fermentation 6–7, 93–4
fire 225
   in forests 152
Fisher, R.A. 60, 199–200
fisheries 156–8
fitness 204–6
flatworms 259
food 103–20
   'natural' 107–9
   tabus 109–10
Forcipomyia 36
forests xii, 151–4, 238–44
   and carbon dioxide 240
   DNA of 55
   in India 248–9
   renewal of 248–9
fossils 13–18
   and human evolution 223–6
fox 167
foxglove 35
Franklin, R. 49
Frisch, K.V. 177
fungi 93, 257
   slime 254
fur tree 28, 54

gardens 147–9, 150
gastrulation 111
gatherer hunters 228
   and diet 108
   populations of 170
   and territory 185
genes 58–86
   engineering the 28, 53–4
   fictional 67, 70, 74–5
   therapy by 54–6
Genesis 23, 24, 237
German measles 41
Gigantopithecus 223
Giganotosaurus 16
giraffe 192, 212

Gombrich, E.H. 3
gonads 103, 118
Gould, S.J. 14
Grassi, G.B. 92
grasslands 14–16
   destruction of 238
great tit 184
greenhouse effect 150, 239, 244
green revolution 247
Griffith, F. 49
gross domestic product 243
gross national product 245
group selection 207
gulls 210

Haldane, J.B.S. 200, 204
Hallucigenia 15
Hebb, D.O. 136
Helicobacter pylori 87–8
Helmholtz, H.L.F. 211
Helmont, J.B. van 6
hemlock 36
hemophilia 62, 64, 68
hemorrhagic fever 45
heredity 58, 66
heritability 75–6
Hess, K. von 177
Hevea brasiliensis 35
HIV 43–6
homeorhesis 110
homeostasis 100
homeothermy 100–2
   of plants 256
'*Homo egoisticus*' 230
Homo erectus 225
homology 21–2
'*Homo pugnax*' 229
Homo sapiens 219–36
   characteristics of 220–2, 231–6
homosexuality 73–4
honey bee 176–9, 256
Hooke, R. 34
Hopi 64

hormones 100–1, 102–3
   of plants 256
hornbill 180
horses 156
   clever 175
   evolution of 21–2
Horton, R. 47
house mouse 167, 206
Hoyle, F. 3
Human Genome Organization (HUGO) 71–3
human immunodeficiency virus (HIV) 43–6
hunter gatherers *see* gatherer hunters

Huntington's disease 72
Huon pines 239
Huxley, J.S. 187
Huxley, T.H. 25
hybridizing 52, 208-9
Hydra 117, 258
hypothalamus 103, 129

Iguanodon 16
imitation 189-91
  by birds 188
immune response 39-42
immunoglobulins 42, 114
imprinting 187-8
induction, embryonic 118-19
influenza 43
Inia geoffrensis 153
instinct 186-8, 236
intelligence 76-8
  and conditional reflexes 133
  of animals 175
intelligence quotient (IQ) 76-8, 80-2
irrigation 240, 247

Jacana spinosa 181
Johnson, S. 109, 234
jumping genes 202-4

Kant, I. 83
Kashamura, A. 43
king cobra 38
Kitasato, S. 166
knee jerk 126, 127
Koch, R. 31
Kozol, J. 251
Krakatau 88-9, 148

Lake Victoria 168-9
Lamarckism 196-8, 204
lamellar cataract 68
language 231-4, 236
Laniarius aethiopicus 181
Larus, species of 210
Lashley, K.S. 131
Lathen, E. 54
Laveran, A. 92
leaf protein 248
learning 124-42 *passim*
  social 198-9
Leeuwenhoek, A. van 31
Leipoa ocellata 179
Lemur catta 182
Leopold, A. 251
leptin 103
leucocytes 33, 40-2
leukemia 33, 119-21
  and radiation 5
  and trade 55

life *see* living things
life style
  and health 73, 76
  and intelligence 106
  and obesity 104
  WHO on 78-9
Linnaeus, C. 19
living things 3
  characteristics of 11
  classified 253-64
  composition of 8
  origin of 8
localization in brain 129-31
locusts 163-5
lymphocytes 40-2
lymphoid organs 40

Macaca 183
McClintock, B. 202-3
machine 12-13
  the organism as 99, 141
McMichael, A.J. 241
malaria 91-3
  and classification 21
  and human numbers 172
  and quinine 36
  and WHO 92-3
mallee fowl 179
mangrove trees 151
Marple, J. 138-40
mathematics 140, 233
Maynard Smith, J. 198
Medawar, P.B. v, 50
meiosis 63
melanism of insects 200-2
melanoma 5, 119
Mendel, G. 59-60
Metchnikoff, I. 40
microscope 31-3
  electron 33
milk 114-15, 170
mitosis 63
moas 167
Mollusca 259-60
moulds 93-5
mushrooms 257
music 234
  of duetting birds 181
mutation 48, 64
  and evolution 208, 209
  and transposons 203
Mycobacterium tuberculosis 45
mycorhiza 256
myxomatosis 162-3

Naja hannah 38
naturalism, biological 228
natural selection 26, 199-202, 211-15

and myxomatosis 163
versus Darwinism 197
'Neanderthal man' 223, 225–6
negative feedback 39
  in cells 37–8
  in populations 160
neuron 124–7
neurosis, experimental 133
niacin 67, 107
Nicotiana tabacum 36
nicotine 36
nitrogen cycle 88, 149–50
nitrogen fixation 53, 88
northern jacana 181
Nyctosaurus 18

oak 63
On the Origin of Species 199
Onychophora 15
Opabinia 14
opium 35
organisms *see* living things
Oryctolagus cuniculus 162–3
osteoporosis 71
Ostrom, J.H. 25

Paley, W. 196
pancreas 102
Pan paniscus 183, 219–20, 232
Pan troglodytes 183, 190, 219–20
Papaver somniferum 35
Papio 182
Parus major 184
Pasteur, L. 6–7, 93, 99
pasteurization 7
Pavlov, I.P. 132–4
peck order 182
pellagra 67, 107
pentadactyl limb 22
peppered moth 200–1
permaculture 247
phagocytes 34, 40–1
phenylketonuria (PKU) 69
pheromones 221
  of bees 177
photosynthesis 35, 37
phrenology 59, 130
pigmy chimpanzee *see* bonobo
Pioneer Foundation 59
Pirie, N.W. 248
pituitary gland 102
plague 47, 165–7
Plasmodium 91–3
pleiotropy 7
plesiosaurs 17
Podocnemis expansa 153
Pogonophora 260
pollution 149–51, 241–2

polyandry 180–1
polygyny 180
populations 159–62
  of fish 156–8
  of people 169–73, 250
  of whales 158–9
Primula 208
prokaryote 9, 13, 87
Protista 90–3
psychosurgery 131
Pteranodon 17
Pterosauria 17
Ptilonorhynchidae 179–80
putrefaction 6, 89
pyloric stenosis 56

Qetzalcoatlus 17
queen (insect) 174, 177
Quercus robur 63
quinine 36

race 79–82
radiation 5
Read, A. 55
Reagan, R. 25
recapitulation 21–3, 112–13
recessiveness 61, 69
red blindness 62, 68
Redefining Progress 243
Redi, F. 6
reduction, explanatory 141–2
red-winged blackbird 184
reflex 126, 127
Reimarus, H.S. 195
respiration 11, 37–8, 93
Rivulus marmoratus 151
RNA 9, 51
Ross, R. 92
rubber tree 35
rubella 41
ruminants 186
Russell, B.A.W. 12

Saccharomyces cerevisiae 37, 159–60
Saiga tartarica 155
Saimiri sciureus 182
sanitary idea 241
Sayers, D.L. 100
schizophrenia 73–4
Schjelderup-Ebbe, T. 182
Schleiden, M.J. 32
Schwann, T. 32
SCID 39
Scombridae 157
'selfish gene' 204, 206–7
severe combined immuno-deficiency
  (SCID) 39
sex differences 82–5

sex linkage 62
sexual reproduction 47–8
Shaw, G.B. 134
sickle-cell anemia 70
signals
   human 188–9
   of animals 176, 187
Silent Spring 242
Skinner, B.F. 134–6
smallpox 41–2
Smith, W. 13
sociobiology
   and human society 230–1
   and sex differences 84–5
soil 149–50
   organisms of 88–9
songs of birds 181
species 208–10
speech 231–4, 236; *see also* language
spiders 261–2
   signals of 176
spontaneous generation *see* abiogenesis
squirrel monkey 182
stabilizing selection 200
statistical analysis 80
stature
   and diet 105–6, 122
   genetics of 80–1
   heritability of 75–6
status system 182–3
Stegosaurus 17
stimulation, need for 136–8
Strachan, T. 55
stress 100
stromatolites 13, 88
substrate 37
succession, plant 148
supersociality of insects 176–9
'survival of the fittest' 199, 204
sweat 104
symbiosis 89, 156, 256
synapse 125
syphilis 44

Tatusia novemcincta 111
Taurotragus oryx 155
Tay-Sachs disease 69, 71
teaching 189, 235
teeth 264
   of birds 22–3
   of *Homo sapiens* 220
   of horses 18
temperature
   of body 100–2
   of environment 4
tendon reflex 126, 127
teratoma 118, 119
termites 174

territory 183–5
thalidomide 122
Theobroma cacao 36
thermometer bird 179
thyroid gland 102
Tinbergen, N. 175–6
tools
   Acheulian 227
   of animals 190
   of early humanity 226–7
   Oldowan 226
transfection 52–4
transgenic organisms 53
transposons 202–4
tuberculosis 31, 45–6
tunnies 157
twins 111
Tyrannosaurus 16

ultradarwinism 213
Ussher, J. 12
Uvarov, B.P. 164

Varanus 16
variola 41–2
vervet monkey 231
viruses 39–40, 41, 43–5
   and cancers 121
   and transposons 203
vitamins 106–7

waggle dance 178
Wallace, A.R. 198–9
warmbloodedness 100–2
warming of biosphere 150, 238–9
water 4, 239
Watson, J.D. 49–51
weeds 168
whales 158–9
Wilkins, M. 49
Wilson, E.O. 231
wine 93–5
Wise Use 242
Wittgenstein, L. 231
wolf 175
women, status of 82–5, 172–3

Xenophanes 12
xeroderma pigmentosum 69
Xiphias gladius 156

Yanomamö 44
yeasts 93–4
   and enzymes 37
   populations of 159–60
Yersin, G.A.E. 166
Yersinia pestis 166